Materials in Extreme Environments

MATERIALS RESEARCH SOCIETY
SYMPOSIUM PROCEEDINGS VOLUME 929

Materials in Extreme Environments

Symposium held April 20–21, 2006, San Francisco, California, U.S.A.

EDITORS:

Daryush ILA
Alabama A&M University
Normal (Huntsville), Alabama, U.S.A.

Christian Mailhiot
Lawrence Livermore National Laboratory
Livermore, California, U.S.A.

Premkumar B. Saganti
Prairie View A&M University
NASA Center for Applied Radiation Research
Prairie View, Texas, U.S.A.

Materials Research Society
Warrendale, Pennsylvania

CAMBRIDGE UNIVERSITY PRESS
Cambridge, New York, Melbourne, Madrid, Cape Town,
Singapore, São Paulo, Delhi, Mexico City

Cambridge University Press
32 Avenue of the Americas, New York NY 10013-2473, USA

Published in the United States of America by Cambridge University Press, New York

www.cambridge.org
Information on this title: www.cambridge.org/9781107408814

Materials Research Society
506 Keystone Drive, Warrendale, PA 15086
http://www.mrs.org

First published 2006
First paperback edition 2012

Single article reprints from this publication are available through
University Microfilms Inc., 300 North Zeeb Road, Ann Arbor, MI 48106

CODEN: MRSPDH

ISBN 978-1-107-40881-4 Paperback

CONTENTS

THEORY AND MODELING
(HIGH PERFORMANCE SIMULATION)

SYNTHESIS AND GROWTH

*Invited Paper

DIAGNOSTICS AND NOVEL CHARACTERIZATION TECHNIQUES

ENVIRONMENTAL EFFECTS
(TEMPERATURE, RADIATION, CORROSION,
EROSION, AND PRESSURE)

*Invited Paper

PREFACE

Symposium II, "Materials in Extreme Environments," was held April 20–21 at the 2006 MRS Spring Meeting in San Francisco, California.

This symposium brought together communities investigating the fundamental properties and response of materials in extreme environments such as static and dynamic high pressure, high strain and high strain-rates, high radiation and electromagnetic fields, high and low temperatures, corrosive conditions, environments causing embrittlement, and environments containing atomic oxygen. The symposium attracted scientists from a broad spectrum of fields of research including space science, planetary science, high-pressure research, shock physics, ultrafast science, and energetic materials research. The investigations of the behavior of materials in extreme environments is an extremely active and vibrant field of research because it is now possible to create in the laboratory conditions of pressure, temperature, and radiation such as those found in, for example, planetary interiors and in space. Moreover, advanced simulation methods, coupled with high-performance computing platforms, now afford predictions — on a first-principles basis — of the properties of materials in extreme environments.

We appreciate the support of Lawrence Livermore National Lab, NASA, Alabama A&M University Research Institute, Alabama A&M University, and the Center for Irradiation of Materials at AAMU, Prairie View A&M University, and all others who provided financial funding for the organization of this symposium. Also, thanks to everyone who contributed to the success of this symposium.

Daryush ILA
Christian Mailhiot
Premkumar B. Saganti

August 2006

MATERIALS RESEARCH SOCIETY SYMPOSIUM PROCEEDINGS

MATERIALS RESEARCH SOCIETY SYMPOSIUM PROCEEDINGS

Prior Materials Research Society Symposium Proceedings available by contacting Materials Research Society

Theory and Modeling
(High Performance Simulation)

Mater. Res. Soc. Symp. Proc. Vol. 929 © 2006 Materials Research Society   0929-II01-01

Radiation Shielding Analysis for Various Materials in the Extreme Jovian Environment

William Atwell
13100 Space Center Blvd., The Boeing Company, Mail Code: HB 2-30, Houston, TX, 77059-3556

ABSTRACT

Earlier particle experiments in the 1970s on Pioneer-10 and -11 and Voyager-1 and -2 provided Jupiter flyby particle data, which were used by Divine and Garrett to develop the first Jupiter trapped radiation environment model. This model was used to establish a baseline radiation effects design limit for the Galileo onboard electronics. Recently, Garrett et al. have developed an updated Galileo Interim Radiation Environment (GIRE) model based on Galileo electron data. In this paper, the GIRE model was utilized to generate trapped proton and electron spectra as a function of Rj (Rj = radius of Jupiter = ~71,400 km). Using these spectra and a high-energy particle transport codes (MCNPX and HZETRN), radiation exposures and dose effects for a variety of shielding materials (Al, polyethylene [PE], and Ta plus several other elemental materials for "Graded-Z" portion of the paper) and thicknesses are presented for the Icy Moon, Europa, Ganymede, and Callisto for several orbital inclinations. In addition, an in-depth discussion and absorbed dose calculations are presented for "Graded-Z" materials and several computer codes were utilized for comparison purposes. We find overall there is generally quite good agreement between the various computer codes utilized in the study: MCNPX (Monte Carlo) vs. HZETRN (deterministic) for slab shielding and the comparison of "Graded-Z" shielding using the CEPXS, MCNPX, NOVICE, and NASA JPL codes. Finally, we conclude that the merits of using "Graded-Z" materials that include PE, due to cost and weight, should aid future Jupiter mission planners and spacecraft designers.

INTRODUCTION

During nearly eight years in the Jupiter magnetosphere, the Galileo spacecraft made a number of major scientific discoveries while making 34 orbits of Jupiter. The mission was truly a marvelous scientific and engineering success. However, the spacecraft was constantly beset with radiation-induced anomalies from the extremely harsh radiation environment that required ground support personnel to expend untold hours at considerable expense in correcting and overcoming the radiation-induced systems problems. Low on attitude control capability and to prevent the spacecraft from colliding with one of Jupiter's Icy Moons (Europa, Ganymede, and Callisto), orbital maneuvers were performed and Galileo impacted Jupiter during its 35th orbit on September 21, 2003.The Jupiter Icy Moons Orbiter (JIMO) program, as originally conceived, is a scientific mission to return to Jupiter and to further extend the discoveries made by Galileo, especially whether the Icy Moons harbor oceans beneath their icy surfaces. Additional information on the JIMO mission is located at the website URL: http://www.jpl.nasa.gov/jimo/.

Unfortunately, funding for the JIMO program was canceled. The previous incarnation, NASA/JPL's Europa Orbiter Project, was cancelled last June because its price tag had exceeded well over a billion dollars. The project was renamed the X2000 Advanced Avionics Project to keep technology development alive until a new mission could be designed.

Shortly thereafter, the National Research Council's Space Studies Board completed a Solar System Exploration Survey (SSES) and issued a report (http://www.nas.edu/ssb/newfrontiersfront.html) entitled, "New Frontiers in the Solar System - An Integrated Exploration Strategy." The report noted that a Europa orbiter mission is still a high priority among planetary scientists, though noted that its price tag (and enabling technologies) is a recurring problem. Nevertheless, the SSES recommended a hypothetical mission, dubbed Europa Geophysical Explorer (EGE) that would fall under a new mission category called "Flagship." Note that, as of this moment, EGE does not exist, but is really only a "placeholder" mission concept until NASA can decide whether to resurrect (and fund) a Europa orbiter.

One of the many challenges facing spacecraft designers is to mitigate the radiation effects to onboard systems. To overcome or minimize this obstacle would be a tremendous cost savings. It is the intent of this paper to assist future spacecraft and systems designers to properly identify the systems and parts that will perform satisfactorily for missions operating in the Jupiter magnetosphere.

In this paper the intense radiation environment, which has been modeled based on trapped proton and electron data obtained during the Voyager 1 & 2 and Pioneer 10 & 11 missions of the 1970's and more recently the Galileo Energetic Particle Detector [EPD][1] data, has been produced in terms of proton and electron spectra for three of the four Icy Moons (Europa, Ganymede, and Callisto) as a function of orbital inclination. The radiation effects to onboard spacecraft systems are strictly related to the amount and type of material shielding the system of interest. Depth dose calculations utilizing both Monte Carlo and deterministic methodologies are presented for a baseline, reference material, aluminum, several other shielding materials, and approaches, such as the use of "Graded-Z" materials. These other approaches include the application of low- and high-Z materials and applying strategies that incorporate multi-disciplinary optimization (MDO) techniques.

RADIATION ENVIRONMENTS

The updated GIRE (Galileo Interim Radiation Electron) model[2] was used to generate both integral and differential trapped electron particle spectra for four orbital inclinations, 0 deg, 30 deg, 60 deg and 90 deg, for the Jupiter Icy Moons: Callisto, Ganymede, and Europa. For Europa, trapped proton spectra were generated in the orbital region of ± 0-28.19 deg. It is worth mentioning that the GIRE model differs from the original Divine-Garrett Jupiter radiation model[3]. The GIRE model includes recent Galileo Energetic Particle Detector[1] electron data in the region L [McIlwain parameter][4] = 8-16 Rj, where Rj = radius of Jupiter (~71,400 km) and the electron and proton energies have been extended to 1000 MeV (1 GeV). Also, the proton environment is the same in both models. The GIRE has a built-in Offset Tilted Dipole magnetic field model option, which was utilized in the particle spectra calculations reported herein.

The particle fluxes also vary as a function of longitude at each Icy Moon. There is approximately a factor of two in particle intensity between 110.8 W longitude and ±90 deg (i.e., 20.8 deg and 200.8 deg W longitude). For the calculations discussed below, the "worst case" 110.8 deg W longitude has been used.

Callisto

Fig. 1 shows the trapped electron integral and differential spectra at Callisto as a function of orbital inclination. The update GIRE was used to generate daily particle spectra where the particle energies range from 0.1 MeV to 1000 MeV (1 GeV).

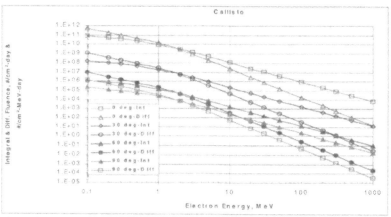

1.   Callisto trapped electron integral and differential spectra using the GIRE model.

Ganymede

Fig. 2 shows the daily, trapped electron integral and differential spectra at Ganymede as a function of orbital inclination.

Europa

Fig. 3 shows the daily-trapped electron integral and differential spectra at Europa as a function of orbital inclination, and Fig. 4 shows the daily-trapped proton integral and differential spectra at Europa as a function of orbital inclination.

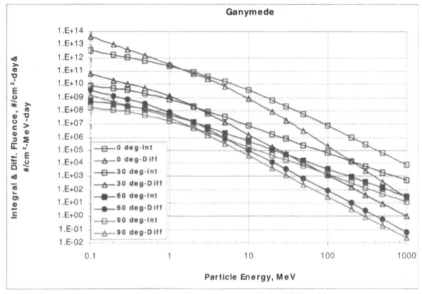

2. Ganymede trapped electron integral and differential spectra using the GIRE model.

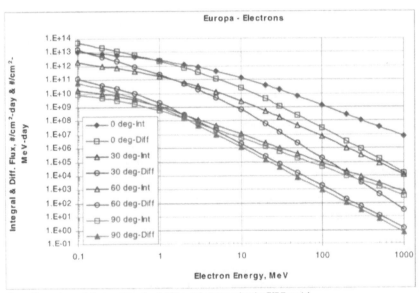

3. Europa trapped electron integral and differential spectra using the GIRE model.

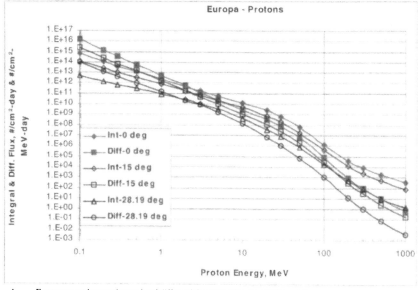

4.   Europa trapped proton integral and differential spectra using the GIRE model.

ABSORBED DOSE CALCULATIONS – MCNPX CODE & ALUMINUM SHIELDING

In this section the absorbed dose calculations are presented using the trapped particle spectra discussed in the previous section.

Callisto electron absorbed doses

Using the MCNPX code[5] and aluminum shielding thickness, the electron absorbed dose rate per day for Callisto is shown as a function of orbital inclination in Fig. 5.

Ganymede electron absorbed doses

Using the MCNPX code and aluminum shielding thickness, the electron absorbed dose rate per day for Ganymede is shown as a function of orbital inclination in Fig. 6. When comparing the electron dose rates for Callisto and Ganymede (Figs. 5-6), we observe almost an order of magnitude dose rate increase at Ganymede, and it is seen in a later section that Europa is ~2 orders of magnitude larger.

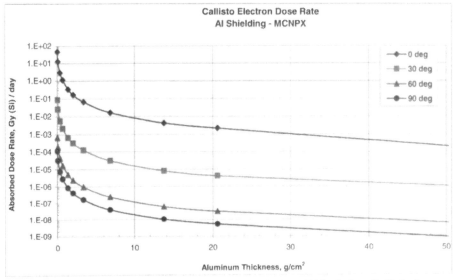

5.   Callisto electron absorbed dose rate using the MCNPX code and aluminum shielding for four orbital inclinations.

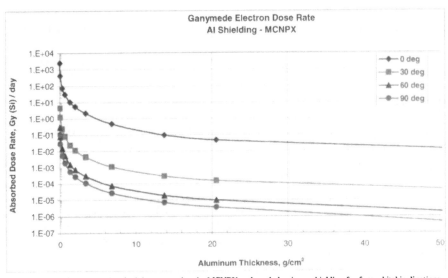

6.   Ganymede electron absorbed dose rate using the MCNPX code and aluminum shielding for four orbital inclinations.

Europa electron and proton absorbed doses

Jupiter's magnetosphere exhibits a particle trapping structure similar to earth, i.e., protons are trapped in an "inner belt" region, and this is such the case at Europa. Figs. 7-8 show the electron and proton absorbed dose rates, respectively, as a function of orbital inclination using the MCNPX code and aluminum shielding. It is noted that the proton trapping extends to approximately ±30 degrees latitude.

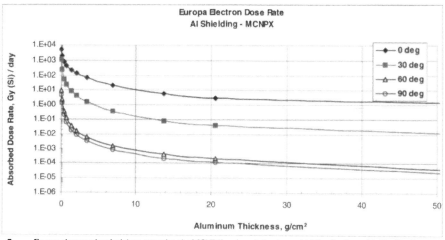

7.   Europa electron absorbed dose rate using the MCNPX code and aluminum shielding for four orbital inclinations.

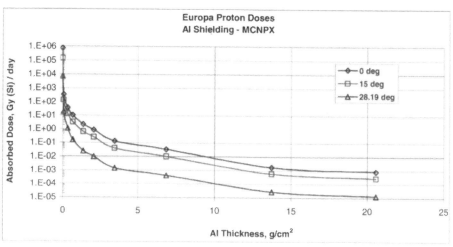

8.   Europa proton absorbed dose rate using the MCNPX code and aluminum shielding for three orbital inclinations.

9

ABSORBED DOSE CALCULATIONS – HZETRN CODE & Al, PE & Ta SHIELDING

In this section the NASA Langley Research Center (LaRC) HZETRN code[6] was utilized to perform similar absorbed dose calculations shown in the previous section, but to include Al, PE and Ta shielding materials. When we compare the MCNPX and HZETRN codes using Al shielding, excellent agreement is observed.

Callisto electron absorbed doses

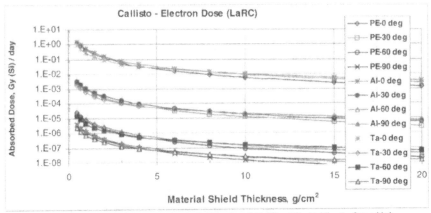

9. Callisto electron absorbed dose rate using the LaRC code and Al, PE and Ta shielding for four orbital inclinations.

Ganymede electron absorbed doses

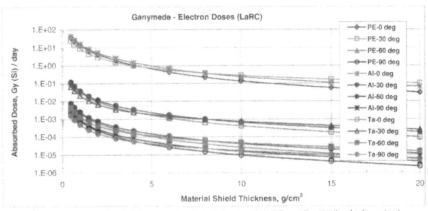

10. Ganymede electron absorbed dose rate using the LaRC code and Al, PE, and Ta shielding for four orbital inclinations.

Europa electron and proton absorbed doses

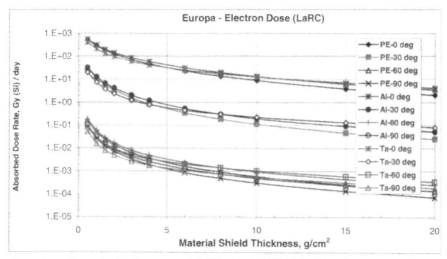

11.   Europa electron absorbed dose rate using the LaRC code and Al, PE, and Ta shielding for four orbital inclinations.

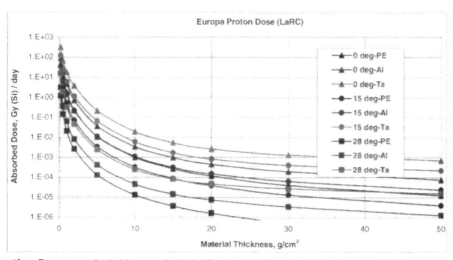

12.   Europa proton absorbed dose rate using the LaRC code and Al, PE, and Ta shielding for three orbital inclinations.

Clearly, from Figs. 11-12, it is observed that at Europa the electrons dominate the dose exposures when compared with the high-energy inner belt proton environment.

GRADED-Z DOSE CALCULATIONS

Fig. 13 is an example of a "Graded-Z" shielding lay-up employing Al and Ta over a Si detector wherein the Europa electron spectrum at 0° inclination is employed and shows the dose attenuation as a function of shielding material using both the CEPXS[7-8] and MCNPX codes. We also compared two other codes, NOVICE[9] and the NASA JPL version of the NOVICE code, although not shown in this paper, and good agreement was also obtained.

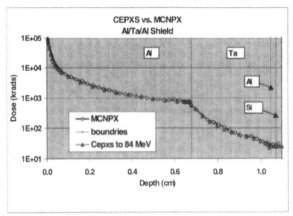

13.   CEPXS and MCNPX absorbed dose comparison for an Al/Ta/Al "Graded-Z" shielding lay-up.

In the strictest sense, the concept of "Graded-Z" shielding was established to utilize various Z-metals to absorb x-rays that are produced as high-energy particles pass through a shielding lay-up. An example of a passive Graded-Z shielding material lay-up would be layered sheets of lead (Pb), tin (Sn) and copper (Cu). Here, Graded-Z shielding refers to the order and atomic number (Z) of those metals (Z equals 82 for lead, 50 for tin, and 29 for copper). Materials with larger Z have greater stopping power, and lead is used for the outermost layer. When a particle hits the lead it will be absorbed and ionize a lead atom, which then emits an X-ray at the "characteristic energy" of lead, 88 keV, in a random direction. Lead does not absorb its own X-rays very well, so to prevent any of those X-rays from getting through the shielding, a layer of tin comes next. After absorbing the 88-keV X-ray, the tin may then emit a 29-keV X-ray. A layer of copper comes last. The few copper X-rays that reach a detector are too low in energy (9 keV) to be of concern.

Table 1 shows specific shield layered thickness(es) that achieve a 25 krad absorbed dose with the least weight for the Europa electron spectrum at 0° inclination case.

Table 1. Shield Layer Thickness(es) to Achieve 25 krad for the Europa Electron Spectrum at 0° Inclination

Shield Material(s)	1st layer (mils)	2nd layer (mils)	3rd layer (mils)
PE	4237		
Al	1517		
Ta / Al	214	10	
Pb / Al	291	20	
U / Al	170	20	
PE / Ta / Al	891	125	10
PE / Pb / Al	816	180	10
PE / U / Al	1540	73	10
Al / Ta / Al	267	145	10
Al / Pb / Al	234	209	10
Al / U / Al	212	124	10

Lastly, Fig. 14 is an example of an iterative optimization technique where both the total shield weight and Ta thickness are minimized to achieve a 50-krad dose for Europa electrons at 0° inclination.

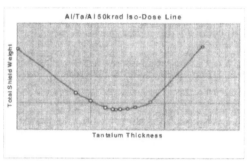

14.   50-krad iso-dose plot of an Al/Ta/Al Shielding to optimize both the total shield weight and Ta thickness.

CONCLUSIONS

The following conclusions can be made:

1. The trapped electron and proton space radiation environments (integral and differential energy spectra) have been generated for the three Icy Moons (Callisto, Ganymede and Europa) of Jupiter based on the GIRE model as a function of orbital inclinationUsing aluminum as a baseline shielding material, absorbed dose calculations as a function of aluminum thickness were generated using the MCNPX code for the three Icy Moons

3. The NASA LaRC HZETRN code was utilized the produce absorbed dose calculations as a function of Al, PE and Ta thickness

4. There was very good absorbed dose agreement between the MCNPX and HZETRN codes using aluminum shielding
5. Using the Europa electron and proton spectra for 0° inclination, the NASA LaRC HZETRN code showed PE was a far superior radiation shielding material when compared with Al and Ta
6. The merits of using Graded-Z lay-up materials showed that optimization techniques can be applied to minimize both weight and material(s) thickness(es) for a given electronics part tolerance threshold
7. The Graded-Z analyses indicated good agreement between the MCNPX and CEPXS codes allowing for the energy-range limitations of the CEPXS code
8. These results should aid in the design of future spacecraft that are to operate successfully and achieve the desired scientific objectives in extreme radiation environments such as those in the magnetosphere of Jupiter
9. The electron and proton spectra and corresponding absorbed dose computations have not been performed for the Icy Moon Io, but the capability exists to do so.
10. Finally, an in-depth discussion of single event upset (SEU) was beyond the scope of this paper. SEU was a major problem for Galileo and onboard electronics systems. However, the Jovian trapped particle environments and materials/shielding analyses presented herein should serve as a guide for mitigating the SEU effects for future missions to Jupiter.

ACKNOWLEDGMENTS

The author gratefully acknowledges the contributions of the following colleagues: Brandon Reddell[1], Bill Bartholet[2], John Nealy[3], Martha Clowdsley[4], Brooke Anderson[4], Thomas Miller[5], and Lawrence W. Townsend[5].

[1]The Boeing Company, Space Exploration, Houston, TX 77059
[2]The Boeing Company, Phantom Works, Seattle, WA 98124
[3]Old Dominion University, Norfolk, VA 23508
[4]NASA Langley Research Center, Hampton, VA 23681
[5]University of Tennessee, Dept. of Nuclear Engineering, Knoxville, TN 37996

REFERENCES

1. D. J. Williams, R. W. McEntire, S. Jaskulek, and B. Wilken, *The Galileo energetic particle detector*, Space Sci. Rev. **60(1-4)**, 385-412 (1992).
2. H. B. Garrett, I. Jun, J. M. Ratliff, R. W. Evans, G. A. Clough, *Galileo interim radiation electron model GIRE*, JPL Publication 03-006 (2003).
3. Divine, N. and Garrett, H. Charged particle distributions in Jupiter's magnetosphere. J. Geophys. Res. **88**(A9), 6889–6903 (1983).
4. C. E. Mc Ilwain, Coordinates for mapping the distribution of magnetically trapped particles, J. Geophys. Res. 66, 3681 (1961).
5. L. S. Waters. (Editor). MCNPX User's Manual, Version 2.4.0. LA-CP-02-408, Los Alamos National Laboratory, Los Alamos, NM (2002).
6. J.W. Wilson, S.Y. Chun, F.F. Badavi, L.W. Townsend, and S.L. Lamkin, HZETRN: A Heavy Ion/Nucleon Transport Code for Space Radiations. NASA TP-3146, NASA Langley Research Center, Hampton, VA, 1991.
7. Lorence, L. J., CEPXS/ONELD Version 2: A discrete ordinates code package for general one-dimensional coupled electron-photon transport, IEEE. Transact. Nucl. Science, Vol. 39, No. 4, p. 1031, 1992.
8. Lorence, L J, et al., Physics guide to CEPX; A multigroup coupled electron-photon cross-section generating code. Sandia National Laboratories Report, SAND-1989, 1989
9. Tom Jordan, Experimental and Mathematical Physics Consultants, Gaithersburg, MD 20885, NOVICE code.

Mater. Res. Soc. Symp. Proc. Vol. 929 © 2006 Materials Research Society 0929-II01-02

Dynamical Fracture Instabilities Due to Local Hyperelasticity at Crack Tips

Markus J. Buehler[1], and Huajian Gao[2]

[1]Civil and Environmental Engineering, Massachusetts Institute of Technology, 77 Mass. Ave., Cambridge, MA, 02139

[2]Division of Engineering, Brown University, Rhode Island, 02912

ABSTRACT

When materials break and cracks propagate, bonds between atoms are broken generating two new material surfaces. Most existing theories of fracture assume a linear elastic stress-strain law. However, the relation between stress and strain in real solids is strongly nonlinear due to large deformation near a moving crack tip, a phenomenon referred to as hyperelasticity or nonlinear elasticity. Cracks moving at low speeds create atomically flat mirror-like surfaces, whereas cracks at higher speeds leave misty and hackly fracture surfaces. This change in fracture surface morphology is a universal phenomenon found in a wide range of different brittle materials, but the underlying physical reason has been debated over an extensive period. Using massively parallel large-scale atomistic simulations employing a new, simple atomistic material model allowing a systematic transition from linear elastic to strongly nonlinear material behaviors, we show that hyperelasticity can play a governing role in dynamical crack tip instabilities in fracture of brittle materials. We report a generalized model that treats the instability problem as a competition between different mechanisms controlled by local stress field and local energy flow near the crack tip. Our results indicate that the fracture instabilities do not only appear in defected materials, but instead are an intrinsic phenomenon of dynamical fracture. Our findings help to explain controversial experimental and computational results, including experimental observation of crack propagation at speeds beyond the shear wave speed in rubber-like materials.

INTRODUCTION

Yoffe's model of a moving crack predicts the occurrence of two symmetric peaks of normal stress on inclined cleavage planes at around 73% of the Rayleigh-wave speed [1, 2]. Gao showed that Yoffe's model is consistent with a criterion of crack kinking into the direction of maximum energy release rate [3]. Marder and Gross presented an analysis including the discreteness of atomic lattice [4, 5], and found instability at a speed similar to other models. Abraham *et al.* [6] suggested that the onset of instability can be understood from the point of view of reduced local

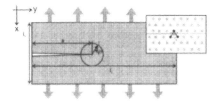

Figure 1: Simulation geometry with crystal orientation and interatomic force-separation laws. The plot shows the simulation geometry with its lattice orientation. We model crack dynamics in a two-dimensional hexagonal crystal.

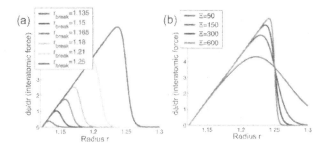

Figure 2: Force versus atomic separation for various choices of the parameters r_{break} and Ξ. Whereas r_{break} is used to tune the cohesive stress in the material, Ξ is used to control the amount of softening close to bond breaking.

lattice vibration frequencies due to softening at the crack tip, and also discussed the onset of the instability in terms of the secant modulus [7]. Heizler *et al.* [8] investigated the crack tip instability based on lattice models [9] using linear stability analysis of the equations of motion including dissipation. They observed a strong dependence of the instability speed as a function of smoothness of the atomic interaction, and pointed out deficiencies in Yoffe's picture. Gao [10, 11] attempted to explain the reduced instability speed based on the concept of hyperelasticity within the framework of nonlinear continuum mechanics. Fracture instabilities have been observed in many experimental studies in many different materials (see, for example [4, 12, 13]).

ATOMISTIC MODELING AND THEORETICAL BACKGROUND

Using large-scale molecular-dynamics (MD) simulations, we model a single crystal under mode I loading with an initial crack as depicted in Fig. 1. Fig. 2 depicts force versus atomic separation of the interatomic potential (symbols r, Ξ, $d\phi/dr$ and r_{break} are defined elsewhere [14]). Figure 2(a) shows the force versus separation curve with respect to changes of r_{break}, and Fig. 2(b) shows the variation in shape when Ξ is varied. For small values of $\Xi \approx 50$, the softening effect is quite large. For large values of $\Xi > 1,000$, the amount of softening close to bond breaking becomes very small, and the solid behaves like one with snapping bonds. The parameter r_{break} allows the cohesive stress σ_{coh} to be varied independently. This model potential describes the limiting cases of material behaviour corresponding to Yoffe's model (linear elasticity with snapping bonds) and Gao's model (strongly nonlinear behaviour near the crack tip). Yoffe's model predicts that the instability speed only depends on the small-strain elasticity. Therefore, the instability speed should remain constant at 73% of the Rayleigh-wave speed, regardless of the choices of the parameters r_{break} and Ξ (thus $v_{inst}^{Yoffe} \approx 0.73 \cdot c_R$). In contrast, Gao's model predicts that the instability speed v_{inst}^{Gao} is *only* dependent on the cohesive stress σ_{coh} and r_{break}:

$$v_{inst}^{Gao} = \sqrt{\frac{\sigma_{coh}}{\rho}} \propto \sqrt{\frac{r_{break}}{\rho}} \tag{1}$$

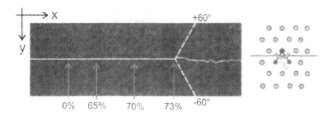

Figure 3: Crack propagation in a homogeneous harmonic solid. When the crack reaches a velocity of about 73 percent of Rayleigh wave speed, the crack becomes unstable in the forward direction and starts to branch (dotted line indicates the 60° plane of maximum hoop stress) [14].

with material density ρ. Variations in Ξ should not influence the crack instability speed. In contrast to previous work focusing on single potential shapes, here we perform numerical studies based on continuously varying potential parameters r_{break} and Ξ.

MODELING RESULTS AND ANALYSIS

We have carried out a series of numerical experiments by systematically varying the potential parameters r_{break} and Ξ. We start with harmonic systems serving as the reference and then increase the strength of the hyperelastic effect. We find that cracks in homogeneous materials with linear elastic properties (achieved by setting Ξ to infinity) show a critical instability speed of about 73% of the Rayleigh-wave speed, independent of the choice of r_{break}, in quantitative agreement with the prediction by Yoffe's model. The crack surface morphology is shown in Fig. 3. We find that in harmonic systems, the occurrence of the instability can be correlated with the development of a bimodal hoop stress as proposed by Yoffe (for example, see [14]).

Once hyperelastic softening is introduced, we observe that the linear elastic Yoffe model fails to describe the instability dynamics. The predictions by Yoffe's model are included in Fig. 4 as the red line, and the predictions by Gao's model are plotted as the blue points. We observe that for any choice of r_{break} and Ξ, the instability speed lies in between the prediction by Gao's model and that by Yoffe's model. Whether it is closer to Gao's model or to Yoffe's model depends on the choice of r_{break} and Ξ. For small values of Ξ and r_{break}, we find that the instability speed depends on the cohesive stress, which is a feature predicted by Gao's model (eq. (1)). We find that the instability speed seems to be limited by the Yoffe speed (see Fig. 4 for $\Xi = 300$ and for large values of r_{break}). Whereas the observed limiting speeds increase with r_{break} for $r_{break} < 1.22$, the results saturate at the Yoffe speed of 73% of Rayleigh-wave speed for $r_{break} \geq 1.22$ (Fig. 4, curve for $\Xi = 300$). In this case, the instability speed is independent of r_{break} and independent of Ξ. This behaviour suggests a change in mechanism, since the instability speed may be governed by a Yoffe-like deformation field mechanism for $r_{break} \geq 1.22$, whereas the instability may be influenced by cohesive-stress and thus energy flow for $r_{break} < 1.22$. We observe a similar transition for different choices of Ξ ranging from about 50 to 1,500, with different transition values of r_{break}.

17

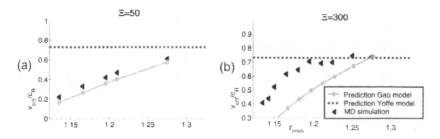

Figure 4: The critical instability speed as a function of the parameter r_{break}, for different choices of Ξ (for $\Xi = 50$ and $\Xi = 300$). The results show that the instability speed varies with r_{break} and thus with the cohesive stress as suggested in [10, 11], but the Yoffe speed [1, 2] provides an upper limit for the instability speed. The instability speeds are normalized with respect to the background strain in the slab, thus including the effect of the slight stiffening of the harmonic lattice with increasing strain.

THE MODIFIED INSTABILITY MODEL

We observe that the first derivative of the instability speed with respect to the cohesive stress in our MD simulations agrees reasonably well with Gao's model. However, the observed instability speed and the prediction differ by a constant value depending on the softening parameter Ξ. We measure the deviation from Gao's model by a shift parameter $v_{shift} = v_{inst}^{MD} - v_{inst}^{Gao}$ which is a function of Ξ (for further details, see [14]). The physical interpretation of v_{shift} is that it accounts for the relative importance of hyperelastic softening close to the crack tip: Gao's model corresponds to the *limiting case* when the softening region is large and completely dominates the energy flow, and it therefore constitutes the *lower limit* for the instability speed. Indeed, we find that for very strong softening ($\Xi \to 0$), the deviation from Gao's model vanishes and $v_{shift} \to 0$. In contrast, the deviation increases to larger values in the case of vanishing softening as $\Xi \to \infty$. The physical significance of v_{shift} can be understood from the perspective of the characteristic energy length scale of dynamic fracture $\chi \propto \gamma E / \sigma^2$ proposed earlier [14, 15]. The characteristic energy length scale χ describes the region from which energy flows to the crack tip to drive its motion. The size of the hyperelastic region is defined as r_H. If the size of the softening region is comparable to χ ($r_H / \chi \gg 1$), hyperelasticity dominates energy flow, and thus $v_{shift} \to 0$, and the predictions of Gao's model should be valid. In contrast, if the size of the softening region is smaller than χ ($r_H / \chi \approx 0$), hyperelasticity plays a reduced role in the instability dynamics and the purely hyperelastic model becomes increasingly approximate, so that v_{shift} takes larger values and eventually Yoffe's model of deformation field induced crack kinking governs. The shift parameter $v_{shift}(\Xi)$ should therefore depend on the relative size of the hyperelastic region compared to the characteristic energy flow

18

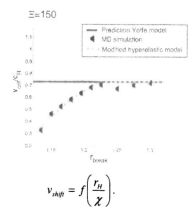

Figure 5. Comparison of the *modified instability model* (see eq. (4)) for $\Xi = 150$ with the MD simulation results, showing the transition from an energy flow controlled instability mechanism to a deformation field controlled mechanism. Our simple model describes the instability dynamics of rapid cracks reasonably well.

$$v_{shift} = f\left(\frac{r_H}{\chi}\right). \tag{2}$$

With the new parameter v_{shift} the instability speed is given by

$$v_{inst} = v_{shift}\left(\frac{r_H}{\chi}\right) + \sqrt{\frac{\sigma_{coh}}{\rho}} . \tag{3}$$

Combing these models, we propose a *modified instability model* in which the critical crack tip instability speed is then given by

$$v_{inst}^{MOD} = \min\left(v_{YOFFE}, v_{shift}\left(\frac{r_H}{\chi}\right) + \sqrt{\frac{\sigma_{coh}}{\rho}}\right) \tag{4}$$

where $v_{YOFFE} \approx 0.73 \cdot c_R$ is a constant, independent on the hyperelastic properties. Fig. 5 compares the predictions by eq. (4) with the results of our MD experiments (dashed curve).

DISCUSSION AND CONCLUSIONS

The work reported here together with earlier results [15] strongly suggest that hyperelasticity, including both nonlinear elastic responses and the geometric aspect of large deformation, is crucial for dynamic fracture, both to understand the instability dynamics as well as to comprehend the crack limiting speed. The onset of instability can be understood as a competition between energy flow governed instability and stress field governed instability (eq. (4)). We hypothesize that the transition between the two mechanisms depends on the relative importance of hyperelasticity around the crack tip, as described by the ratio of the size of the hyperelastic region and the characteristic energy length scale r_H / χ [15]. In most experiments [4, 13] and computer simulations [6, 16], materials show a significant softening effect, explaining the reduced instability speed. Other studies carried out in stiffening materials such as rubber suggest that cracks can even approach intersonic velocities, while still propagating stable and creating atomically flat surfaces. Further details can be found in [14].

ACKNOWLEDGMENTS

The simulations were carried out at the Max Planck Society Supercomputer Center in Munich and at the MARS Linux cluster at the Max Planck Institute for Metals Research in Stuttgart. We gratefully acknowledge their support. MJB acknowledges support from MIT's CEE Department.

REFERENCES

1. Freund LB: *Dynamic Fracture Mechanics.* Cambridge University Press, ISBN 0-521-30330-3; 1990.
2. Yoffe EH: **The moving Griffith crack.** *Phil Mag* 1951, **42:**739-750.
3. Gao H: **Surface roughening and branching instabilities in dynamic fracture.** *J Mech Phys Solids* 1993, **41:**457-486.
4. Fineberg J, Gross SP, Marder M, Swinney HL: **Instability and dynamic fracture.** *Phys Rev Lett* 1991, **67:**457-460.
5. Marder M, Gross S: **Origin of crack tip instabilities.** *J Mech Phys Solids* 1995, **43:**1-48.
6. Abraham FF, Brodbeck D, Rafey RA, Rudge WE: **Instability dynamics of fracture: A computer simulation investigation.** *Phys Rev Lett* 1994, **73:**272-275.
7. Abraham FF: **Unstable crack motion is predictable.** *Advances in Physics* 2005, **53:**1071-1078.
8. Heizler SI, Kessler DA, Levine H: **Mode I fracture in a nonlinear lattice with viscoelastic forces.** *Phys Rev E* 2002, **6:**016126.
9. Slepyan LI: *Models and Phenomena in Fracture Mechanics.* Springer, Berlin; 2002.
10. Gao H: **A theory of local limiting speed in dynamic fracture.** *J Mech Phys Solids* 1996, **44:**1453-1474.
11. Gao H: **Elastic waves in a hyperelastic solid near its plane-strain equibiaxial cohesive limit.** *Philosphical Magazine Letters* 1997, **76:**307-314.
12. Cramer T, Wanner A, Gumbsch P: **Energy dissipation and path instabilities in dynamic fracture of silicon single crystals.** *Phys Rev Lett* 2000, **85:**788-791.
13. Fineberg J, Gross SP, Marder M, Swinney HL: **Instability in the propagation of fast cracks.** *Phys Rev B* 1992, **45:**5146-5154.
14. Buehler MJ, Gao H: **Dynamical fracture instabilities due to local hyperelasticity at crack tips.** *Nature* 2006, **439:**307-310.
15. Buehler MJ, Abraham FF, Gao H: **Hyperelasticity governs dynamic fracture at a critical length scale.** *Nature* 2003, **426:**141-146.
16. Abraham FF, Brodbeck D, Rudge WE, Xu X: **A Molecular Dynamics Investigation of Rapid Fracture Mechanics.** *J Mech Phys Solids* 1997, **45:**1595-1619.

Mater. Res. Soc. Symp. Proc. Vol. 929 © 2006 Materials Research Society 0929-II01-08

Kinetics of the Nucleation and Growth of Helium Bubbles in bcc Iron

Chaitanya Suresh Deo[1], Srinivasan G. Srivilliputhur[1], Michael Baskes[1], Stuart Maloy[1], Michael James[1], Maria Okuniewski[2], and James Stubbins[2]
[1]Los Alamos National Laboratory, Los Alamos, NM, 87545
[2]University of Illinois, Urbana, IL, 61801

Abstract

Microstructural defects are introduced in materials upon irradiation with energetic particles. These defects can cause degradation of mechanical properties and contribute to material failure. Transmuted helium in irradiated stainless steels exerts deleterious effects on material properties. We have performed kinetic Monte Carlo (kMC) simulations of point defect diffusion and clustering in bcc alpha iron. The model includes helium and vacancy diffusion and spontaneous clustering and dissociation of the point defects from the clusters. We employ the kMC simulations to investigate the time evolution of the point defect configuration leading to defect clustering and bubble formation. The concentration of embryonic point defect clusters is determined as a function of the simulation time.

Introduction

During high energy proton irradiation, high-energy neutrons collide with the atoms in the surrounding materials and induce (n, α)-reactions resulting in the formation of helium atoms. Consequently, the first-wall materials in the fusion reactor (typically ferritic steels) contain a high concentration of helium atoms during and after irradiation [1, 2]. These helium atoms have a strong tendency to precipitate into helium-vacancy clusters and bubbles, which are detrimental to the properties of metals and alloys. Studies have shown that helium atoms assist the nucleation and growth of cavities in irradiated materials leading to volumetric swelling. Mechanical properties such as tensile strength and fracture toughness [3, 4] are influenced by the presence of helium atoms. Helium migration and clustering at grain boundaries results in high temperature embrittlement [5]. Thus, understanding helium behavior in metals is key to developing structural materials capable of operation in a high energy proton irradiation environment.

The helium- vacancy cluster evolution under irradiation is governed by several mechanisms responsible for transport of He atoms and vacancies in the crystal, such as the migrating He interstitial, migrating vacancy, thermally activated dissociation of helium from a vacancy and the jump of a He atom from one to another vacancy as a basic step in the vacancy mechanism [6]. The effect of irradiation on materials microstructure and properties is an inherently multiscale phenomenon. Radiation damage processes including helium behavior occur over multiple length and time scales, from the collision stage of 10^{-13} s and 10^{-9} m to the long-term diffusion stage of 10^3 s and 10^6 m. Primary defect production occurs at nanometer and picosecond scales where helium is produced via the (n,α) reaction and vacancies and self-interstitials via the elastic collisions of primary knock on atoms. The long-term diffusion of vacancy, self-interstitial defects and impurity gas atoms is responsible for microstructural evolution. It is generally believed [6, 7] that diffusing helium rapidly clusters to form an evolving population of bubbles, both in the bulk and at grain boundaries. These bubbles act as biased sinks for point defect fluxes and the absorption of helium and/or vacancies. Subsequent evolution results in their conversion to growing voids. In spite of this qualitative understanding, the effect of helium on the evolving micro-structural features in advanced ferritic steels cannot yet be quantitatively predicted.

In this paper, we will focus on a lattice based kinetic Monte Carlo (kMC) code to simulate the long-term diffusion of vacancies and helium in displacement cascades and the modification of defect evolution by the presence of helium in body centered cubic (bcc) iron. The kinetic Monte Carlo method[8], unlike molecular dynamics, is an event-driven technique, i.e., it simulates events at random with probabilities according to the corresponding event rates. In this way, it self-adjusts the time-step as the simulation proceeds, depending on the fastest event present at that time. These kMC methods[9-11] use information on the energetic barriers to defect migration in order to obtain the relationships between defect diffusivity and clustering. Previous kMC simulations have studied the recombination of vacancies and self interstitial atoms in iron and copper. Our kMC model incorporates the migration energies of the point defects (such as self interstitial atoms, vacancy, interstitial and substitutional helium and hydrogen), cluster formation energies, dissociation energies of the point defects from the clusters. The KMC simulations predict the evolution of the vacancy and helium concentration as well as the concentration and composition of small embryonic helium vacancy clusters.

Simulation Model and Algorithm

The present kinetic Monte Carlo (kMC) simulations study the migration of point defects and the subsequent evolution of helium-vacancy bubbles. The model is capable of simulating the following entities in bcc iron: interstitial and substitutional helium and hydrogen atoms, interstitial iron atoms, vacancies, vacancy-clusters, and sinks for the trapping of point defects (dislocations and grain boundaries). Input to the simulations may include the migration energies of the point defects (interstitial iron, vacancy, interstitial and substitutional helium and hydrogen), formation energies of the He_nV_m clusters, dissociation energies of the point defects

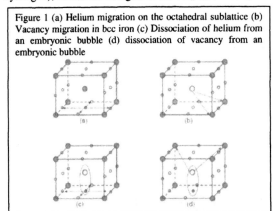

Figure 1 (a) Helium migration on the octahedral sublattice (b) Vacancy migration in bcc iron (c) Dissociation of helium from an embryonic bubble (d) dissociation of vacancy from an embryonic bubble

from the He_nV_m clusters and initial concentrations and configurations of point defects and defect ratios. These energetic parameters can be obtained from molecular dynamics (MD) simulations using empirical potentials or from first principles calculations. Similarly defect ratios and configurations can be obtained from the post-cascade data of large MD runs.

Figure 1 shows the basic mechanisms of helium and vacancy activity in single crystal bcc iron. Large filled circles represent iron, large open circles represent vacancies, small filled circles represent helium atoms and small open circles represent the octahedral bcc sites. The bcc iron lattice and the interstitial octahedral lattice can be collectively represented as a simple cubic lattice in three-dimensional space as demonstrated in the figure. The lattice parameter of the simple cubic lattice site is $a_0/2$, where a_0 is the lattice parameter of bcc iron. The stable configuration of helium interstitials (octahedral or tetrahedral) depends on the inter-atomic potential employed to describe the iron-helium system. Potentials based on the Finnis Sinclair method [12] suggest the octahedral sites as the lower energy sites. Recent first principles calculations [13] suggest that the lowest energy configuration might be the tetrahedral site. It is generally agreed that the difference in energy between the helium interstitial at a tetrahedral site and one at the octahedral site in bcc iron is very small. It is possible to include both interstitial sites in our kMC simulations. In the present paper, we assume the helium to occupy the octahedral position on the interstitial sub-lattice.

The kMC model used in this paper consists of helium interstitials on the octahedral sublattice and vacancies on the bcc iron lattice. The rates of migration of the point defect entities are calculated as

$$r^i_{migration} = v^i_{migration} \exp\left(-\frac{E^i_{migration}}{k_B T}\right),$$
(1)

where the superscript i refers to the helium and the vacancy point defect entities. The rate of migration of the point defect entity is $r^i_{migration}$, the attempt frequency is $v^i_{migration}$, the migration barrier is $E^i_{migration}$, while k_B and T are the Boltzmann constant and the temperature respectively. Two point defect entities are considered to be in a cluster when the distance between them is less than a_0, which is the lattice constant of bcc iron. The rate of dissociation of a point defect entity (i = helium or vacancy) from a cluster into the bulk lattice is considered to be thermally activated and is calculated as:

$$r^i_{dissociation} = v^i_{dissociation} \exp\left(-\frac{E^i_{dissociation}}{k_B T}\right)$$
(2)

where $r^i_{dissociation}$ is the rate or dissociation, $v^i_{dissociation}$ i the attempt frequencies, $E^i_{dissociation}$ is the energy of dissociation. The attempt frequencies $v^i_{dissocation}$ and $v^i_{migration}$ are taken to be equal to the Debye frequency of bcc iron. The dissociation energy $E^i_{dissociation}$ of a point defect from a cluster is taken to be the sum of the energy to bind a point defect entity to the cluster and $E^i_{migration}$. Morishita et al [12] have calculated the migration energies of helium and vacancies as well as the binding energies of some helium-vacancy clusters.

Table I shows the Kinetic Monte Carlo event catalog. The migration and dissociation events of helium and vacancy point defect entities are parameterized by the energy expressed in eV.

Point Defect Entity	Event	Lattice from	Lattice To	Direction	Energy (eV)
Helium	Migration	Octahedral	Octahedral	<100> 4 out of 6	0.078
Vacancy	Migration	Substitutional	Substitutional	<111>	0.9
Helium	Dissociation from $He_n V_m$	Substitutional	Octahedral	<100> 4 out of 6	2.0
Vacancy	Dissociation from $He_n V_m$	Substitutional	Substitutional	<111>	1.90
Helium	Dissociation from He_n	Octahedral	Octahedral	<100> 4 out of 6	0.30
Vacancy	Dissociation from V_m	BCC substitutional	BCC substitutional	<111>	1.10

Table I shows the energetic parameters employed to describe the diffusion of the helium and the vacancies and the dissociation energies of helium from He_n and He_eV_m clusters and vacancies from a V_m and He_nV_m cluster. These energy parameters are used to calculate the rates of all the possible events (Equations 1-2) in the system and build the event catalog for the kinetic Monte Carlo simulation.

Simulations are performed with the helium and vacancy point defect entities randomly distributed in the simulation cell. The helium atoms occupy interstitial sites while the vacancies are distributed by substitution in the bcc iron lattice. The event catalog is generated by calculating the rates of migration or dissociation of the point defect entities using Equations (1) and (2). The kMC event catalog consists of the migration, clustering and dissociation of two point defect entities, helium and vacancies. The transition probability of each event is proportional to the rate of event occurrence, calculated by the Equations (1) and (2). We use the standard kMC algorithm[8, 13, 14] to select an event and update simulation time with a variable time increment.

At each kMC step, the system is monitored to identify a clustering event. When two point defect entities (helium-helium, vacancy-vacancy, helium-vacancy) are in a cluster the simulation creates a mapping between the entities and the cluster such that for each cluster there are at least two entities associated with the cluster. The event catalog is updated with the new rates of event occurrence and the transition probabilities for the next kMC event are calculated using Equations (1-4). The concentrations and compositions of He_nV_m clusters are recorded at each kMC step. In this paper, we report on the time evolution of the concentration of vacancies and helium in the bulk, helium in bulk and cluster, helium (He_n) clusters, helium-vacancy (He_nV_m) clusters.

Results

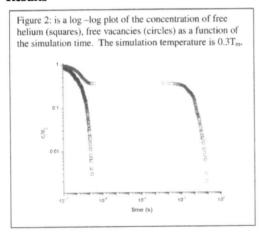

Figure 2: is a log –log plot of the concentration of free helium (squares), free vacancies (circles) as a function of the simulation time. The simulation temperature is $0.3T_m$.

We employ the above kMC model and algorithm to study helium and vacancy diffusion and clustering over small and large time scales. The interstitials diffuse by hopping to adjacent octahedral sites on the lattice while the vacancies diffuse by the vacancy mechanism by substitution in the bcc iron lattice. It is possible to introduce He atoms and vacancy populations throughout the kMC simulation at rates appropriate to the irradiation environment. In the present note, we introduce 1000 helium interstitials and 1000 vacancies into the simulation cell at randomly determined bcc octahedral sites and bcc substitutional sites respectively. The simulation size is taken to be 120x120x120 unit cells. If the helium interstitial is in close vicinity (less than a_0, the unit cell parameter of the bcc iron lattice) of another helium interstitial, the two atoms are considered to be in a cluster. The procedure described in the previous section is repeated for 10^6 kMC steps and the defect concentrations are monitored throughout the simulation.

In Figure 2, we plot the evolution of vacancy and helium concentration as a function of the kMC simulation time. Both free helium and free vacancy concentrations decrease with increasing time. The free helium concentration decreases far more rapidly than the free vacancy concentration. After the free helium concentration decreases to a negligible amount, the free vacancy concentration decreases rapidly. The concentration of helium-helium clusters increases in the initial evolution phase, but decreases to a negligible amount at longer times. At small time scales (less than 1 nanosecond), helium atoms rapidly diffuse on interstitial sites and get trapped at the relatively slow moving vacancies. At higher time scales, vacancy mobility is higher and this leads to coalescence of the vacancy-helium clusters.

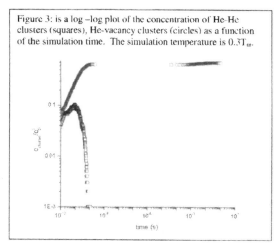

Figure 3: is a log –log plot of the concentration of He-He clusters (squares), He-vacancy clusters (circles) as a function of the simulation time. The simulation temperature is $0.3T_m$.

Figure 3 shows the evolution of the helium vacancy (He_nV_m, squares) and the helium-helium (He_n, circles) clusters. The ppm concentration of He_nV_m and the He_n clusters is plotted as a function of simulation time. The concentration of helium-helium clusters increases in the initial evolution phase, but decreases to a negligible amount at longer times. The concentration of helium-vacancy (He-V) clusters increases throughout the simulation and reached a constant value once all the free helium and the vacancies are incorporated in bubbles. At low times (less that 1ns), the relatively fast moving helium atoms cluster together or get trapped at the slow moving vacancies. This leads to the initial increase in concentration of He_nV_m and He_n clusters. The helium atoms then dissociate from the embryonic He_n clusters leading to a decrease in the He_n concentration. Finally after 10^{-5} seconds, vacancy diffusion becomes more pronounced, leading to vacancy coalescence and an increase in He_nV_m concentration. The He_n clusters are less stable than the helium-vacancy clusters; hence their concentration decreases as more helium is trapped at vacancy sites.

In the initial stages, the gas atoms diffuse through the lattice and cluster with each other as well as with vacancies. Both gas-V and gas-gas dissociation can occur in the system, however, the gas-gas dissociation energy is far smaller. The dissociation of the gas atoms from gas-gas clusters is thermodynamically more favorable than dissociation from gas-V clusters. Thus gas-V clusters persist at steady state while the gas-gas clusters dissolve with a short lifetime. Gas-gas clusters persist longer when the vacancy concentrations are low as the gas atoms collide with each other more often. The dissolution of the gas-gas clusters agrees with experiments of radiation damaged iron with implanted helium[21] that observe that helium atoms are mostly present in the form of gas bubbles rather than in isolated interstitial helium clusters

Summary

We have employed the kMC simulations to investigate the time evolution of the point defect configuration leading to defect clustering and bubble formation. The concentration and composition of embryonic defect clusters as a function of time and operating temperatures is determined. The low migration energy of the helium interstitials implies that the helium diffusion is faster than the vacancy diffusion. This leads to the initial formation of He_nV_m and He_n clusters. These cluster concentrations evolve with time as the vacancy migration becomes more active at a higher time scale. Future work will clarify the mobility of vacancy-He cluster complexes and extend the simulation durations to predict He bubble nucleation. This approach may also be appropriate for the simulation of hydrogen in condensed matter systems.

Acknowledgements

This work is performed under the auspices of the U.S. Department of Energy and Los Alamos National Laboratory and the Advanced Fuel Cycle Initiative.

References

[1] F. Garner, B. Oliver, et al., Journal of Nuclear Materials, 4th International Workshop on Spallation Matereials Technology (IWSMT-4) **296**, 66-82 (2001).
[2] B. M. Oliver, M. R. James, et al., Journal of Nuclear Materials, 10th International Conference on Fusion Reactor Materials (ICFRM-10) **307**, 1471-1477 (2002).
[3] S. Maloy, M. James, et al., JOURNAL OF NUCLEAR MATERIALS, 5th International Workshop on Spallation Materials Technology **318**, 283-291 (2003).
[4] S. A. Maloy, M. R. James, et al., in 3rd Workshop on Utilisation and Reliability of High Power Proton Accelerators edited by OECD, 105-124, (2003)
[5] T. R. Armstrong and P. J. Goodhew, Radiation Effects **77**, 35-48 (1983).
[6] H. Trinkaus and B. N. Singh, Journal of Nuclear Materials, **323**, 229-42 (2003).
[7] M. BASKES, R. FASTENAU, et al., Journal of Nuclear Materials **102**, 235-245 (1981).
[8] A. B. BORTZ, M. H. KALOS, et al., Journal of Computational Physics **17**, 10-18 (1975).
[9] H. L. Heinisch and B. N. Singh, Philosophical Magazine, **83**, 3661-76 (2003).
[10] H. L. Heinisch, B. N. Singh, et al., Journal of Nuclear Materials, **276**, 59-64 (2000).
[11] C. Domain, C. S. Becquart, et al., Journal of Nuclear Materials **335**, 121-145 (2004).
[12] K. Morishita, R. Sugano, et al., Nuclear Instruments & Methods in Physics Research Section B-Beam Interactions With Materials and Atoms, **202**, 76-81 (2003).
[13] C. C. Battaile and D. J. Srolovitz, Annual Review of Materials Research **32**, 297-319 (2002).
[14] K. A. FICHTHORN and W. H. WEINBERG, Journal of Chemical Physics **95**, 1090-6 (1991).

Mater. Res. Soc. Symp. Proc. Vol. 929 © 2006 Materials Research Society 0929-II01-10

Experimental and Ab-Initio Investigations of Osmium Diboride

M. M. Hebbache[1], L. Stuparevic[2], D. Zivkovic[2], and M. Zemzemi[1]

[1]Materiaux et Phenomenes Quantiques (UMR 7162), Universite Paris 7, 2 Place Jussieu, Paris Cedex 05, F-75251, France

[2]Department of Metallurgy, University of Belgrade, VJ12, Bor, 19210, Yugoslavia

Abstract

More than half a century after their discovery, almost nothing is known about the physical properties of osmium borides, though their structures have been clearly identified in the early sixties. We re-examined the phase diagram of the binary system osmium-boron and confirm the existence of two hexagonal phases, $OsB_{1.1}$, Os_2B_3 and an orthorhombic phase, OsB_2 Our microhardness measurements show that the synthesized OsB_2 is extremely hard. In addition, first-principles calculations have been conducted to investigate its physical properties. It is shown that OsB_2 is also a low compressibility material. Most of the transition metal borides have already found applications as in protective armor, nuclear reactors, reinforcement, etc. OsB_2 can be used in applications such as hard coating.

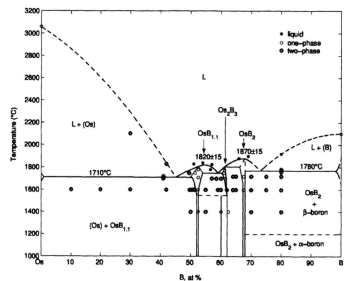

Figure 1: Phase diagram of Os-B system, with OsB_{11} (hex.), Os_2B_3 (hex., high and low temperature modification) and OsB_2 phases, respectively.

1 Introduction

Technological applications of superhard materials are countless, e.g., protective armor, nuclear reactors, space-flight, turbine technology, etc. Up to date, light atoms like carbon, boron and nitrogen are considered as the best candidates to form superhard materials. Recently, it has been found that osmium, which is the heaviest elemental crystal, is less compressible than diamond [1]. Low compressibility materials are often superhard. Consequently, there is an increased interest in this transition metal and its compounds. Several experimental and theoretical works have been already devoted to osmium. For instance, it has been predicted recently that osmium could undergo reconstructive and isostructural phase transitions at very high pressures [2, 3]. As for numerous transition metal carbides, nitrides and borides, it is expected that binary compounds of osmium with light elements could also be superhard [4]. However, to our best knowledge, osmium nitride does not exist in crystalline form while osmium carbide OsC is either metastable or has a narrow temperature-stability range [5]. On the contrary, three osmium borides have been synthesized in the early fifties [6]. A decade later, their structures have been clearly identified by two different groups [7, 8].

Though the properties of transition metal borides and their applications have been intensively investigated, almost nothing is known about osmium borides. The main reason for such data lack is that very high temperatures, larger than 1800 °C, are required to synthesize these materials. Moreover, osmium still has bad press, due to the formation of the tetroxide $OsO4$ which boils at 130 °C and which is extremely toxic. We recently overcame these difficulties and confirmed the composition and the structure of the three borides $OsB_{1.1}$, Os_2B_3 and OsB_2 [9]. In the present work, we give an extensive study of OsB_2 which includes metallography, hardness measurements and ab-initio calculations of its physical properties.

2 Synthesis and Structure

The synthesis of osmium borides is not an easy task. Powder mixtures of high purity (> 99.5 %) osmium and boron were pressed into pellets (mass of 2 g) and then arc-melted under purified (99.9999 %) argon atmosphere in a water-cooled crucible. To insure a sufficient homogeneity, samples were cut and remelted several times. A lot of knowledge and experience were needed to prepare correctly the samples. The Os-B phase diagram is shown on figure (1). Metallography was used for the determination of phases and their regions [9]. OsB2 crystallizes in the RuB2-type structure with an orthorhombic cell containing two chemical formula [7, 8]. The unit cell and the structure of the metal and boron layers are shown on figure 2. The space group number is 59 (Pmmn). The lattice parameters a, b and c are 4.684, 2.872 and 4.076 Å, respectively. The two osmium atoms occupy the set (a) : (1/4,1/4,z) and (3/4,3/4,z) with z = 0.1535 while the four boron atoms occupy the set (f): (x,1/4,z)), (x+1/2,1/4,z), (x,3/4, z) and (x+1/2, 3/4, z) with x = 0.058 and z = 0.632. Each boron atom is surrounded by three boron atoms and four metal neighbours while each osmium atom has eight boron atom neighbours. The structure has voids at the set 2(b) with z = 0.8. These voids are sufficiently large to accommodate two B atoms. All atoms have a fixed y-coordinate, 1/4 or 3/4, forming dense atomic planes along the b-parameter. Along the c-direction, a description of this structure can be based on atomic layers. In comparison with the AlB_2 structure, in which a large group of metal (Me) diborides with chemical formula MeB_2 crystallize, OsB_2 exhibits a deformed two-dimensional network of boron atoms. The latter form corrugated hexagons with four B – B – B angles at 106.84° and two at 99.97° instead of six angles at 120°. The boron layer lies between two planar metal layers which are offset (see top of Fig.2).

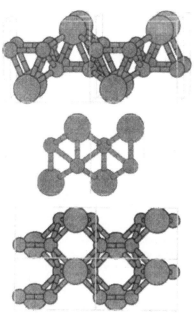

Figure 2: Top : Crystal structure of OsB_2 showing 2x2x1 unit cells. There are two formula per unit cell. The vertical and the horizontal directions are along c- and a-parameters, respectively. Osmium and boron atoms are represented by large (grey) and small (green) circles, respectively. The corrugated boron layer is between two planar metal layers which are offset. Projection along b-axis can be readily drawn from this perspective. Middle : projection of the OsB_2 structure along a-axis. The horizontal direction is along b-parameter. All atoms have a fixed position y = 1/4 or 3/4; forming parallel atomic layers along b-parameter. Bottom : projection of the OsB_2 structure along c-axis. The vertical and the horizontal directions are along b- and a-parameters, respectively. Boron atoms are at the vertices of corrugated hexagons.

3 Hardness

It is well known that transition metal borides exhibit a high hardness [4]. Their Vickers hardness is in the range of 2000-3000 kg/mm^2. We have performed microhardness measurements on the (001) plane of OsB_2 by means of the PMT-3 tester. Using a diamond pyramidal indenter, the Vickers hardness was calculated from the relation $H_V = 1854.4\ p/d^2$, where p is the applied load and d the length of the diagonal of the indentation impression. With p in gram units and d in μm, H_V turns out to be expressed in kg/mm^2 units. The applied load gives a stress distribution $\sigma_{zz}(x, y)$ beneath the indenter. x and y are the components of a two-dimensional vector lying on the free surface of the sample z = 0 (for a detailed study of the indentation problem, see Ref. [10]). The Vickers hardness of OsB_2 is about 3600 kg/mm^2 for p in the range of 60-80 g. For higher loads, the hardness decreases significantly, e.g, 3000 kg/mm^2 at 100 g. A Vickers hardness greater than 3000 kg/mm^2 has been quoted in a recent patent [11]. At higher loads, the hardness depends strongly on crack formation. OsB_2 is evidently among the most hard transition-metal compounds [4].

4 Electronic properties

The microhardness is governed by the strongest bonds which prevent the close approach of the atoms under compression. In what follows, we investigate the electronic, elastic and plastic properties of OsB_2 in order to have a deeper understanding on bonding. Our computational investigation was carried out within the density functional theory [12]. We used WIENk2k [13] and FHI98MD [14] packages. The method that WIEN2k uses is known as the full-potential linearized augmented-plane-wave (FP-LAPW). We employed the generalized gradient approximation (GGA) for the exchange-correlation energy [15]. The $[Kr]4d^{10}$ states of osmium and the $1s^2$ state of boron were treated as core states. As a convergence criterion, the energy difference between two successive iterations, was used within a value of $0.2 \mu Ry$. Such a high accuracy level is required for the computation of the elastic properties (see below). It was achieved by considering a k-mesh of 1000 points in the Brillouin zone (140 in the irreducible part) and a large basis set cut-off, i.e., $R_{mt}^{min}K_{max} = 11$. The muffin tin radius

R_{mt} of boron was equal to half of the B-B internuclear distance while that of Os was fixed such that the atomic spheres are almost touching. We used the FHI98MD package to supply some limitations of WIEN2k. This code uses a norm-conserving pseudopotential to describe the potential of the nuclei and cores electrons. The valence electron wavefunctions were expanded in plane waves with the kinetic energy cutoff E_{cut} of 130 Ry. The pseudopotential of Hamann type [16] in the fully separable form of Kleinman and Bylander [17], was employed. The pseudopotentials were generated with the $5d^6 6s^2$ and $2s^2 2p^1$ atomic configurations of osmium and boron, respectively. The GGA as parameterized by Perdew et al. [15] has been again used. For the Brillouin zone sampling we used a 6x6x6 Monkhorst-Pack mesh [18]. Electronic properties are characterized by energy bands, density of states (DOS) and valence electronic charge density (VECD). The band diagram along various directions connecting high-symmetry points of the Brillouin zone is shown on figure 3. It is clear that OsB_2 has a metallic character. The DOS and the integrated DOS (dashed curve) are also displayed in figure 3. There is a large valence electron concentration, i.e., 28 valence electrons per unit cell. The partial DOS in Os and B depend on the radius of the atomic spheres R_{mt} and are not given here. The total DOS is composed of a large d-component of Os and small sp-components of Os and B. The contribution of the f-electrons of Os is negligible. In the orthorhombic structure, the d orbitals split into four irreducible representations, namely, $2A_g$, B_{1g}, B_{2g} and B_{1g}. The bonding energy band is mainly due to the hybrid state of 5d and sp-boron orbitals while the anti-bonding states are extended states. The Fermi level lies in a DOS minimum. This confers a relatively high stability to the orthorhombic structure of OsB_2. The charge density profiles in three different planes where the three bondings Os-Os, B-B and B-Os are present, are shown in Fig.4. The inter-atomic distances which play an important role are given below. These profiles reveal that a large amount of valence electron charge is localized at the osmium atoms (large peaks). This is in agreement with the DOS results. A cross-section of a plane containing four osmium atoms at the vertices of a lozenge is shown in the top of Fig.4. In this plane, the Os-Os inter-atomic distances are the shortest, i.e., 2.872 and 3.018 Å. The low value of the VECD between atoms suggests that the covalent component of Os-Os bonds is not significant.

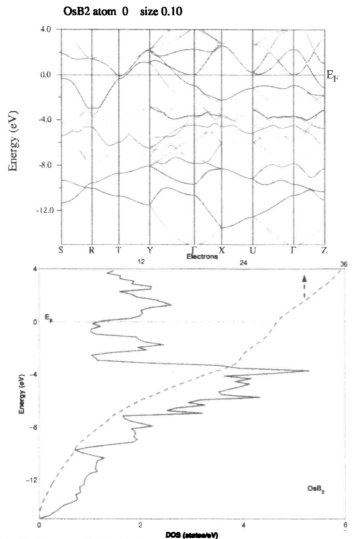

Figure 3: Top : Band structure of OsB$_2$ along high symmetry directions. Zero of the energy scale corresponds to the Fermi energy. Bottom : Density of states of OsB$_2$ (solid curve) and integrated density of states (dashed curve).

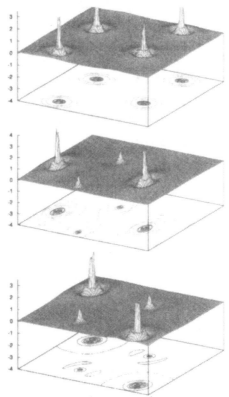

Figure 4: Valence electron charge density (VECD), in atomic units, and contour plots. Large peaks are due to electrons localized at atomic sites. Top : VECD in a plane containing four osmium atoms at the vertices of a lozenge. Middle : VECD in the dense plane y = 1/4 or 3/4. Bottom : VECD in the incline which links the planes y = 1/4 and 3/4. Inter-atomic distances are given in the text.

strain	deformation-gradient matrix	Fuchs constants
v	$J_{kk} = v^{1/3}$	$[C_{11} + C_{22} + C_{33} + 2(C_{12} + C_{13} + C_{23})]/9$ $+ C_{22} + C_{33} + 2(C_{12} + C_{13} + C_{23})]/9$
ε	$J_{11} = (1 + \varepsilon)^{1/2}, J_{22} = 1/J_{11}, J_{33} = 1$	$(C_{11} + C_{22} - 2C_{12})/4$
γ	$J_{kk} = 1, J_{23} = \gamma$	C_{44}
εv	$J_{11} = (1 + \varepsilon)^{1/2} v^{1/3}, J_{22} = v^{2/3}/J_{11}, J_{33} = v^{1/3}$	$(C_{11} - C_{22} + C_{13} - C_{23})/6$

Table 1: Fuchs strains, non-zero components of the deformation-gradient matrices J and the corresponding Fuchs constants expressed by means of Brugger-type elastic constants C_{IJ}

On the left hand side of figure 4, the VECD is shown in a cross-section of either plane y = 1/4 or 3/4. In these y-planes, the boron atoms are covalently bonded. The inter-atomic B-B distance, 1.798 Å., is the shortest one. In the incline which links the y-planes (see Fig.2), the inter-atomic B-B distance is slightly larger, i.e., 1.874 Å.. The boron atoms are less bonded in the incline as shown on the right hand side of Fig.4. The inter-atomic distances between boron and metal atoms are larger than the B-B inter-atomic distances, i.e., 2.147 and 2.308 Å. in y-planes and 2.215 Å. in the incline. A significant amount of valence electron charge lies between boron and osmium atoms. The Os-B bonds are also due to the delocalization of the metal electrons. In short, addition of boron leads to the increase of the valence electron concentration and the formation of directional bonds which strengthen the structure of OsB_2 anisotropically.

5 Elastic properties

The material hardness is intimately connected to the strength of the chemical bonds. The latter are well illustrated by force constants in valence-force-fields models. Calculations of harmonic and anharmonic force constants can be done for simple crystalline structures [19]. However, for more complex structures, the algebra is tedious. Moreover, force constants cannot be measured. It is then simpler to consider elastic constants for representing the chemical bond strengths. In principle, they are connected to force constants [19]. Elastic constants can be measured by means of Brillouin spectroscopy and ultrasonic techniques. The hardness is correlated to bulk and shear moduli, which express the material resistance to volume and shape changes, respectively [20]. These coefficients can be expressed by means of Brugger-type elastic constants C_{IJ}, defined as the second derivatives of the energy per unit undeformed volume with respect to the components of the Lagrangian finite strain tensor $\eta = (J*J - I)/2$. I is the unit matrix and J the deformation-gradient matrix whose effect is to transform any position vector of a point in the unstrained lattice x to x' = $Jx+u(n)$ where u(n) is the internal strain or inter-lattice displacement of the nth sublattice. The nine independent elastic constants of OsB_2 can be calculated by imposing small deformations of the unit cell and polynomial fits to the total energy. It is computationally simpler to use infinitesimal strains similar to those introduced by Fuchs [21] and then the resulting constants converted to Brugger constants C_{IJ}, their finite strain counterparts. We considered a first distortion of Fuchs-type v which expands or contracts the three axes uniformly. A second deformation ε is chosen so that one axis is expanded and a second axis contracted. The third strain γ tilts one axis and leaves unchanged the base formed by the pair of orthogonal axes. The two latter distortions are volume-conserving. The first two deformations are symmetry-conserving while the third distortion transforms the orthorhombic lattice into a monoclinic lattice. The corresponding deformation-gradient matrices J_{ij} are given in Table 2. By a cyclic permutation of indices, one get the nine Fuchs constants which allow to calculate the nine Brugger-type elastic constants C_{IJ}. The latter, obtained for distortions less than 0.05, are given in Table 2 with that of the two hardest materials, i.e., diamond and cubic boron nitride c-BN. The value of C_{33}, 875 GPa, is larger than that of c-BN, and almost comparable to that of diamond, 1079 GPa. This is qualitatively in agreement with the hardness measurements because the elastic constant C_{33} expresses mainly the material resistance to the relative change of the c-parameter. The lower values of C_{11} and C_{22} suggest that the hardness will be reduced by one third along a and b-axes, i.e., 2400 kg/mm^2. A strong anisotropy in hardness has also been observed in TiB2 and is usually taken into account in applications such as coating [4]. The bulk modulus B can be calculated by

33

considering the volume strain v or deduced from the values of C_{IJ} (see Table 2). Applying the second method, we found that B is

Material	C_{11}	C_{22}	C_{33}	C_{12}	C_{13}	C_{23}	C_{44}	C_{55}	C_{66}	B	Ref.
OsB$_2$	611	597	875	204	157	133	279	137	342		This work
c-BN			820			190		480	400		[23]
diamond			1076			125		577	442		[24]

Table 2: Brugger-type elastic constants and bulk modulus, in unit of GPa, of osmium diboride and that of the two hardest materials, i.e., diamond and cubic boron nitride.

equal to 342 GPa. Our result is about 13 % smaller than that obtained by fitting the measured pressure-vs-volume curve [11]. Osmium diboride is then a low compressibility material. Actually, B is not the best choice to estimate the hardness because of the strong material anisotropy of OsB2. For metallic materials, the hardness is mainly associated with the ease to generate and move dislocations by indentation. The shape of the indenter and the orientation of the primary slip systems are of major importance. An orthorhombic crystal possesses only one slip plane, here y-plane, and two slip directions, x and y-axes [22]. These two primary slip systems are associated with the shear moduli C_{66} and C_{44}. The latter express, respectively, the material resistance to shape change under the shear stresses σ_{xy} and σ_{zy}. The theoretical value of C_{44} cannot be obtained with accuracy because the orthorhombic lattice of OsB2 is unstable to the shear strain ~ (see Table 2). An energy maximum is reached for a low value of γ, i.e., 0.006, the starting point of the lattice instability. The monoclinic phase is stable at γ equal to 0.04, the value for which an energy minimum is obtained. The calculated value of C_{66} is low (see Table 3). This is in agreement with the charge density profiles and hardness measurements. The chemical bonds in the incline which links the y-planes, 1/4 and 3/4, are more sensitive to shear forces than to compressive forces.

6 Conclusion

Several decades after three independent groups [6, 7, 8], we succeeded in synthetizing three osmium borides, i.e., OsB$_{1.1}$, Os$_2$B$_3$ and OsB$_2$. Hardness measurements and first-principles calculations have been conducted to investigate electronic, elastic and plastic properties of osmium diboride OsB$_2$. We found that it is a low compressibility material and a hard material, in particular along the c-axis. Comparison of pure osmium which crystallizes in the hcp structure and OsB$_2$ shows that the addition of the three-dimensional network of boron atoms decreases the compressibility by 10 to 20 % while the hardness is multiplied by a factor of 5 to 9. Such hardness anisotropy should be taken into account in applications such as hard coating.

References

[1] Cynn, H., Klepeis, J. E., Yoo, C. S., Young, D. A., *Phys. Rev. Lett.* 88, 135701-4 (2002).

[2] Occelli, F., Farber, D. L., Badro, J., Aracne, C. M., David, D. M. and Hanfland, M., *Phy. Rev. Lett.* 93, 095502 (2004).

[3] Hebbache M. and Zemzemi, M., *Phys. Rev. B* 70, 224107-6 (2004).

[4] Holleck, H., *J. Vac. Sci. Technol. A* 4, 2661-2669 (1986).

[5] Kempter, C. P., *J. Chem. Phys.* 41, 1515-1516 (1964).

[6] Buddery, J. H. and Welch, A. J. E., *Nature* 167, 362-362 (1951).

[7] Aronsson, B., Stenberg, E. and Aselius, J., *Nature* 195, 377-378 (1962); Aronsson, B., *Acta Chem. Scand.* 17, 2036 (1963).

[8] Roof, R. B., Jr. and Kempter, C. P., *J. Chem. Phys.* 37, 1473-1476 (1962).

[9] Stuparevič, L. and Živkovič, D., *J. Therm. Analys. Calor.* 76, 975-983 (2004).

[10] Willis, J. R., *J. Mech. Phys. Solids* 14, 163-176 (1966).

[11] R. B. Kaner, J. J. Gilman and S. H. Tolbert, *Science* 308, 1268-1269 (2005); J. Am. Chem. Soc. 127, 7264 (2005).

[12] Kohn, W. and Sham, L. J., *Phys. Rev.* 140, A1133-1138 (1965)

[13] Blaha, P., Schwarz, S., Madsen, G. K. H., Kvasnicka, D. and Luitz, J. *Computer Code WIEN2k* (Vienna University of Technology, Vienna, 2001).

[14] Bockstedte, M., Kley, A., Neugebauer, J. and Scheffler, M., *Comput. Phys. Commun.* 107, 187-222 (1997).

[15] Perdew, J. P., Burke, K. and Ernzerhof, M., *Phys. Rev. Lett.* 77, 3865-3868 (1996).

[16] Bachelet, G. B., Hamann D. R. and Schluter, M., *Phys. Rev. B* 26, 4199-4228 (1982).

[17] Kleinman, L. & Bylander, D. M., *Phys. Rev. Lett.* 48, 1425-1428 (1982).

[18] Monkhorst, H. J. & Pack, J. D., *Phys. Rev. B* 13, 5188-5192 (1976).

[19] Vanderbilt, D., Taole, S. H. and Narasimhan, S., *Phys. Rev. B* 40, 5657-5668 (1989).

[20] Teter, D. M., *Mater. Res. Soc. Bull.* 23, 22-27 (1998).

[21] Fuchs, K. A., *Proc. Roy. Soc. A* 153, 622-639 (1936).

[22] Hebbache, M. *Solid State Commun.* 113, 427-432 (2000).

[23] Grimsditch, M., Zouboulis, E. S. and Polain, A., *J. Appl. Phys.* 76, 832-834 (1994).

[24] Grimsditch, M. and Ramdas, A. K., *Phys. Rev. B* 11, 3139-3148 (1975).

[25] Shackelford, J. F. and Alexander, W. *CRC Materials Science and Engineering Handbook* (Third ed., CRC Press, USA, 2006)

Mater. Res. Soc. Symp. Proc. Vol. 929 © 2006 Materials Research Society   0929-II04-20

Atomistic Studies of Crack Branching at Bimaterial Interfaces: Preliminary Results

Sriram Krishnan[1], and Markus J Buehler[2]

[1]Department of Mechanical Engineering, Massachusetts Institute of Technology, 77 Mass. Ave, Cambridge, MA, 02139

[2]Department of Civil and Environmental Engineering, Massachusetts Institute of Technology, 77 Mass. Ave, Room 1-272, Cambridge, MA, 02139

ABSTRACT

In this paper we summarize recent progress in applying atomistic studies of cracking along interfaces of dissimilar materials under quasi-static crack growth conditions. We consider two linear-elastic material strips in which atoms interact with harmonic potentials, with a different spring constant in each layer leading to a soft and a stiff strip. The two strips are bound together with a tunable potential, which allows to independently control the interface and bulk fracture surface energy . An initial crack serves as initiation point for the failure. This provides a model system to investigate how elastic properties and interface strength interplay and determine the crack growth direction, leading to either interfacial cracking or branching into the film material. We observe a clear transition to interface failure when the interface fracture energy is less than 80% of the bulk fracture energy. We further find that branching in the film material is controlled by the elastic properties of the film material, suggesting interfacial cracking for extremely soft films and branching for stiffer films. Analysis of the virial stress field around the crack suggests that the circumferential hoop stress controls the branching behavior.

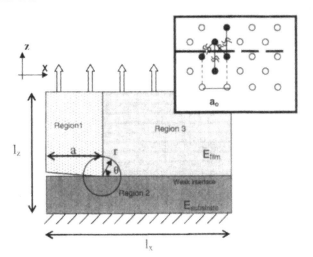

Figure 1: Simulation geometry and lattice orientation for the studies of interfacial cracking. The figure shows a stiff substrate with a soft film on top, under mode I (tensile) loading applied to the film. The simulation domain is dividing into several regions. The interface between region 1 and 2 serves as a starter crack. The schematic further shows the definition of crack angle around the crack tip. The inlay shows the orientation of the atomic FCC lattice in [100][010][001] orientation.

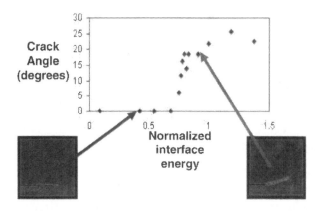

Figure 2. The angle at which crack grows in region 3 is shown, as a function of the ratio of interface fracture energy to bulk fracture energy. When the interface fracture energy is up to 80% of the bulk fracture energy, the crack angle is $0°$ and the crack grows along the interface. Beyond the 80% limit, the crack has a sharp change in behavior and grows at an angle of about $20°$ in the bulk.

INTRODUCTION

Atomistic simulation is becoming an increasingly important tool to investigate fundamental aspects of crack initiation and propagation. Recent progress in this field include systematic atomistic-continuum studies of fracture [1,2], investigations of the role of hyperelasticity in dynamic fracture [3] and studies on the instability dynamics [4]. There have been several continuum-based studies on stress analysis of bi-material wedges and the order of stress singularity around the crack tip [5-8]. Recent work using MD simulations on polymer films have shown that equilibrium interface energies alone can not be used to predict interface failure behavior [9]. Here we focus on linking microscopic potential parameters that lead to changes in elastic properties and surface energies to the macroscopic branching behavior of cracks, with the goal to develop a physical generally valid understanding of which parameters control crack initiation and branching at interfacial cracks. We extend existing atomistic models of fracture in homogeneous materials to study cracking along interfaces of dissimilar materials. Such investigations are motivated by both the scientific and technological relevance of the general problem of bimaterial interfaces.

The most important research objective of our studies is the branching behavior of cracks: Using molecular-dynamics, can we determine what is the criterion that governs crack angle at an interface between dissimilar materials? How does branching change for soft interfaces? Such studies are relevant to understand the deformation mode of cracks along material interfaces, as present in organic films – for example soft biological matter – deposited on stiff substrates. An immediate technological application is peeling of polymer films in soft lithography and other industrial applications such as prevention of biofouling and in the design of adhesives.

Interatomic potentials for a variety of different brittle materials exist, many of which are derived form first principles. However, it is difficult to identify generic relationships between potential parameters and macroscopic observables such as the crack limiting or instability speeds when using such complicated potentials. We deliberately avoid these complexities by adopting a simple pair potential based on a harmonic interatomic potential. To obtain a clean model of cracks at interfaces, we study two half spaces with harmonic interatomic potentials, but with different spring constants $k_2 < k_1$. The ratio $\Xi = k_2 / k_1$ measures the elastic mismatch of the two

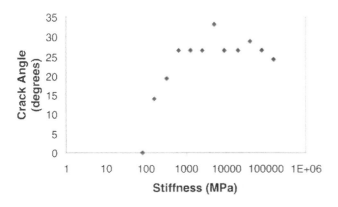

Figure 3. The crack angle in region 3 is shown in a linear-log plot, as a function of the stiffness in region 3. The stiffness parameter in the simulation, k_3 varies from 0.5 to 1000, corresponding to an actual stiffness value in the film of about 80 MPa to about 160 GPa. For the same ratio of film to substrate stiffness, the crack angle varies vastly depending on the stiffness value.

materials. The wave speeds are different by a factor $\sqrt{\Xi}$. The harmonic potential is given by

$$\phi(r) = \frac{1}{2}k_i(r - r_0)^2 H(r_{break,ij} - r),$$ (1)

where r_0 is the atomic equilibrium spacing, r denotes the atomic distance, k_i is the stiffness parameter of the harmonic potential in region i, $r_{break,ij}$ is the threshold breaking distance for bonds between regions i and j, and H is the heaviside function accounting for bonds breaking when $r > r_{break,ij}$. Young's modulus is given by $E_i = 2k_i / \sqrt{3}$ and the shear modulus is $\mu_i = \sqrt{3}k_i / 4$.

The system under study is shown in Figure 1. Atoms in Region 1 do not interact with atoms in region 2, leading to an initial crack of length a. We are interested in understanding the fracture behavior around the crack tip shown. The inlay in Figure 1 shows the FCC atomic lattice structure used in the simulation. The stiffness ratio between region 3 and region 2 is 1:10, which is held constant in all simulations. Our simulations simulate a softer material on a stiff substrate. For comparison, PMMA on Al has a stiffness ratio about 1:15. We can specifically weaken the interface by changing the value of $r_{break,23}$ which defines the bond breaking distance between region 2 and 3. The sample is loaded under mode I plane strain conditions.

The outline is as follows. We start with a parametric study to investigate (i) the effect of relative interface strength on branching, and (ii) the effect of the stiffness of the film material. These studies are realized by systematically varying the stiffness parameter k_i and the threshold breaking radius $r_{break,ij}$. We then investigate the which criteria influences the crack branching angle, including the local hoop stress distribution or the angular variation of the fracture surface energy. All studies are carried out under remote tensile (mode I) loading. Our results suggest that a peak in hoop stress ahead of the crack determines the direction of initial crack propagation.

CRACK BRANCHING BASED ON RELATIVE INTERFACE STRENGTH

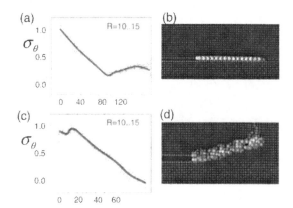

Figure 4. Subplot (a), hoop stress around crack tip (reference value=160 MPa). Subplot (b), snapshot showing crack direction. For interface crack, the hoop stress profile decreases sharply starting at an angle of $0°$. Subplot (c), plot of hoop stress around crack tip (reference value=160MPa). Subplot (d), snapshot showing crack direction. For crack in the bulk of the film, the hoop stress profile is nearly flat up to an angle of about $20°$, which is nearly equal to the crack angle. The red line is drawn to guide the eye.

In the first set of simulations we vary the strength of the interface and investigate the fracture behavior. To weaken the interface, we reduce the critical length for bond breakage $r_{break,23}$ between region 2 and 3 (Figure 1). The interfacial fracture surface energy γ_{23} (for surface creation between regions 2 and 3) is given by

$$\gamma_{23} = N_b \rho \frac{1}{4} k_i (r_{break,23} - r_0)^2, \tag{2}$$

as a function of $r_{break,23}$, where $r_0 = 1.12246\text{Å}$. We vary $r_{break,23}$ from 1.14 Å to 1.19 Å and observe the fracture behavior in the system. The breaking radius for the bonds in region 3 (film) is 1.18 Å and the breaking radius of bonds in region 2 (substrate) is 1.18 Å. The substrate is specifically made stronger – by choosing a larger breaking radius – so that it does not alter the fracture behavior between region 3 and region 2. If we use a lower value of $r_{break,22}$ in region 2, we notice more complicated failure behavior where region 2 also undergoes a symmetric crack growth in the middle. When we normalize the interface fracture energy (equation 2) relative to the fracture energy in the film (region 3), we find that failure occurs at the interface upto a ratio of 80 %. Figure 2 depicts the initial crack angle as a function of the normalized interface energy. For interface fracture energy higher than 80%, there is a sharp transition into failure in the film. When $r_{break,23} = 1.18\text{Å}$, the normalized interface fracture energy is 1. Further increases in $r_{break,23}$ beyond 1.18 Å does not show any interface fracture behavior or of appreciable change in crack angle.

CRACK BRANCHING BASED ON STIFFNESS IN THE FILM

Figure 3 shows a study of the influence of film stiffness in region 3 on the branching angle. The stiffness parameter of the film, k_i is varied over 3 orders of magnitude from 0.5 to 1000, corresponding correspond to a Young's modulus range of about 80 MPa to 160 GPa. In all cases, the stiffness of the substrate (region 2) was maintained at 10 times the stiffness of the film (region 3). Figure 3 shows the dependence of the crack angle on k_3 in a linear-log plot. We

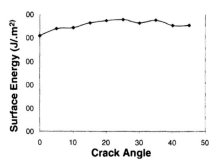

Figure 5. Variation of surface energy with angular position, in degrees, around the crack tip. Surface energy at the crack tip increases initially for increasing crack angle, reaching a maximum value near a crack angle of 20~25° and then decreases. The relative variation in surface energy for various crack angles is a small fraction (<20%) of the absolute value of surface energy.

measure the crack angle at the crack tip by considering the location of the fractured surface at a horizontal distance of 10 atoms from the crack tip. We observe that for small values of film stiffness, the failure occurs at the interface. For higher values of k_3, there is a varied dependence of the crack angle on the stiffness. This observation will need further study as traditional fracture mechanics predicts crack angle at a bimaterial interface to be constant for a fixed E_3 / E_2 ratio.

CRACK BRANCHING BASED ON HOOP STRESS

Here we focus on finding a possible criterion of what may control the direction of crack growth. As suggested in studies related to crack instabilities and branching, changes in local stress field may control the crack growth direction [1, 3]. Figure 4 show the study of crack branching based on hoop stress profile σ_θ (circumferential stress) around the crack tip. We plot hoop stress in a band of atoms at a radial location of 10 to 15 atoms spacing from the crack tip. For each atom in the radial band, we obtain the cylindrical coordinate positions relative to the crack tip. The stress tensor at each atom in the band is then transformed using the Mohr's circle to obtain the hoop stress value at the atom. For different interface strength values, the hoop stress profile gives indications regarding the nature of crack growth (in the interface versus in the bulk).

Figure 4(a-b) shows the hoop stress profile when the crack grows along the interface. The hoop stress profile in the case of interface crack shows a steep profile with the maximum value at 0°. Figure 4(c-d) shows the hoop stress profile when the crack branches into region 3. The hoop stress profile shows a plateau in the range of 0-20° and the crack angle is nearly 20°. These results indicate that the change in hoop stress may control the branching angle.

CRACK BRANCHING BASED ON SURFACE ENERGY VS ANGLE

Figure 5 shows the variation of surface energy with respect to angular position around the crack tip. In this simulation, planar cracks originating at the crack tip at pre-determined angles were introduced in region 3. The total energy of the system was computed. Based on the reference energy of the system, we then obtain the work done in separating along the angular plane. By computing the surface area introduced by the planar cracks, we can estimate the fracture surface energy at a particular angle around the crack tip. We observe that the fracture surface energy varies with the angular position around the crack tip from 0° to about 20°, stays nearly constant and then decreases for angle higher than 35°. The variations in the surface energy with angle are a small fraction of the absolute value of surface energy, suggesting that they play a minor role in controlling the branching direction.

DISCUSSION AND CONCLUSIONS

The branching property of cracks at interfaces is significantly influenced by the material properties at the interface, and the properties of the materials surrounding the interface. We showed that interfacial fracture is preferred up to an interface strength of 80%. For interfaces with strengths beyond 80% of the film strength, cracks tend to branch off into the film material.

We further find that for constant interface strength and variations in elasticity, crack branching is preferred for stiff materials. For extremely soft materials, cracks tend to grow along the interface. An analysis of the stress field surrounding an interfacial crack reveals that crack branching is controlled by the hoop stress: Broad hoop stress distributions lead to crack branching, whereas shallow peaks of the hoop stress lead to interfacial fracture. Preliminary analysis of the fracture energy in various directions suggests that the fracture energy variation with respect to angular changes could play a minor role, corroborating our hoop stress based concept. The studies carried out here were done for simplistic interatomic potentials, within the framework of computational experiments. Even though these results do not represent a real material, the main findings could be generally valid in real materials.

ACKNOWLEDGMENTS

Joe Foley assisted in developing scripts for the simulations and we acknowledge his contribution. MJB acknowledges support from MIT's CEE Department.

REFERENCES

1. Gao, H.J., Huang, Y., Abraham, F.F. Continuum and Atomistic Studies of Intersonic Crack Propagation, J. Mech. Phys. Solids **49**, 2113-2132 (2001).
2. M.J. Buehler, F.F. Abraham, H. Gao, Hyperelasticity Governs Dynamic Fracture at a Critical Length Scale, Nature **49**, 441-446 (2003)
3. M.J. Buehler, H. Gao, Y. Huang, Continuum and Atomistic Studies of Suddenly Stopping Supersonic Cracks, Computational Materials Science **28**, 385-408 (2003).
4. Abraham, F.F., Brodbeck, D., Rudge, Instability Dynamics of Fracture: A Computer Simulation Investigation W.E., Xu, X. Phys. Rev. Lett. **73**, 272-275 (1994).
5. Rice, J.R., Sih, G.C. Plane Problems of Cracks in Dissimilar Media. Trans. of the ASME **32**(2), 418-423 (1965)
6. England, A.H. A Crack Between Dissimilar Media. J. Appl. Mech. 32, 400-402 (1965)
7. Rice, J.R. Elastic fracture mechanics concepts for interfacial cracks. Trans. of the ASME **55**(1), 98-103 (1988)
8. Williams, M.L. The stresses around a fault or crack in dissimilar media. Bull. Seismol. Soc. America 49, 199-204 (1959).
9. Gersappe, D., Robbins, M. O., Where do polymer adhesives fail?. Europhys. Lett, **48** (2), pp. 150-155 (1999).

Mater. Res. Soc. Symp. Proc. Vol. 929 © 2006 Materials Research Society   0929-II04-21

Effect of Pressure on Electronic Structure of Pb$_{1-x}$Sn$_x$Te Alloys Doped with Gallium

Evgeny Skipetrov[1], Alexander Golubev[2], Nikolay Dmitriev[1], and Vasily Slyn'ko[3]

[1]Faculty of Physics, M.V.Lomonosov Moscow State University, Moscow, 119992, Russian Federation

[2]Faculty of Material Sciences, M.V.Lomonosov Moscow State University, Moscow, 119992, Russian Federation

[3]Institute of Material Science Problems, Chernovtsy, 274001, Ukraine

ABSTRACT

The galvanomagnetic effects in the n-Pb$_{1-x}$Sn$_x$Te:Ga (x=0.09-0.21) alloys at the temperatures 4.2≤T≤300 K and under hydrostatic compression up to 16 kbar have been investigated. It is shown that in all samples and in the whole investigated pressure range temperature dependencies of resistivity and Hall coefficient have a "metallic" character, indicating stabilization of Fermi level by the impurity resonant level. Using the experimental data in the frame of two-band dispersion law the dependencies of the free electron concentration and the Fermi level position upon temperature, matrix composition and pressure were calculated. The temperature, composition and pressure coefficients of gallium resonant level movement were obtained and the electronic structure under varying the alloy composition and under pressure were built.

INTRODUCTION

Doping of lead telluride based alloys with mixed valence impurities (In, Ga, Tl, Cr, Gd, Yb, etc.) leads to appearance of deep impurity levels stabilizing the Fermi level in the gap or on the background of allowed bands [1, 2]. In PbTe among the dopants mentioned above gallium is the sole impurity that induces the deep level situated within the gap, slightly above the middle of the gap. Hence gallium doping makes it possible to realize a dielectric state with extremely low, close to intrinsic values of the free charge carrier concentration in the fairly narrow range of impurity concentration (C$_{Ga}$=0.2-0.7 mol. %). At the same time there are some experimental data indicating existence of the second gallium-induced defect level lying on the background of conduction band and stabilizing the Fermi level at higher gallium concentrations [3-7]. In particular, our recent studies of Pb$_{1-x}$Ge$_x$Te:Ga (0.04≤x≤0.08) shows that there are at least two different impurity levels E$_{Ga1}$ and E$_{Ga}$ in the energy spectrum of charge carriers in these alloys [6, 7]. The second impurity level E$_{Ga}$ also situates in the gap under the conduction band bottom L$_6^-$ of investigated alloys. The energy position of this level depends on the germanium concentration by the linear law: E$_{Ga}$-L$_6^-$ ≈ 45-1980x meV. So, we have supposed that in PbTe this level should be resonant level situated in the conduction band and lying approximately on 45 meV higher than its bottom. However, there are still no direct experimental evidences of the existence of this gallium resonant level in the conduction band of PbTe and PbTe-based alloys.

This research was devoted to the investigation of the effect of alloy composition and hydrostatic compression on the galvanomagnetic properties of the Pb$_{1-x}$Sn$_x$Te alloys doped with gallium. The main aims were to reveal resonant gallium-induced level E$_{Ga}$ in the conduction band and to obtain the diagrams of electronic structure reconstruction under variation of matrix composition and under pressure. It was supposed that the position of resonant gallium-induced level with respect to the conduction band bottom could be altered by varying the tin content in the alloys and applying pressure, inducing the redistribution of electrons between localized and band states and changing of the free carrier concentration in the alloys.

EXPERIMENTAL DETAILS

Single crystals of n-$Pb_{1-x}Sn_xTe$:Ga ($0.09 \leq x \leq 0.21$, C_{Ga}=0.2–2.2 mol. %) were grown by the modified Bridgman method. The tin ratio and impurity concentration in the doped samples were determined from the initial amount of substance in the furnace charge, taking into account the distribution of impurity along the ingot during the growth process, according to the exponential law of impurity distribution in A^4B^6 solid solutions established in [8], and controlled by the energy dispersive X-ray fluorescence analysis. Pressures up to 16 kbar were obtained in the beryllium bronze chamber with kerosene-oil-pentane pressure transmitting media and were determined at helium temperatures by measuring the superconducting transition temperature in pure tin as a function of pressure in the chamber. For each sample and each value of pressure the temperature dependencies of the resistivity ρ and the Hall coefficient R_H ($B \leq 0.1$ T, $4.2 \leq T \leq 300$ K) were measured by four-probe technique. Table I summarizes the main parameters of the samples at T=4.2 K and atmospheric pressure.

Table I. Parameters of the investigated n-$Pb_{1-x}Sn_xTe$:Ga samples at T=4.2 K.

x	C_{Ga} (mol.%)	ρ (10^{-4} Ω cm)	$-R_H$ (10^{-1} cm^3/C)	n (10^{19} cm^{-3})
0.09	0.2	1.3	9.7	0.65
0.10	0.3	1.4	9.0	0.70
0.11	0.9	1.4	9.4	0.69
0.14	0.5	1.6	8.0	0.79
0.16	0.3	2.7	5.2	1.2
0.18	0.7	1.1	4.1	1.6
0.21	2.2	1.5	2.7	2.3

EXPERIMENTAL RESULTS

It was found that in all samples and in the whole investigated pressure range the temperature dependencies of resistivity ρ and Hall coefficient R_H have a "metallic" character. However, the Hall coefficient is changed in anomalous manner: an absolute value of R_H increases more than two times with increasing temperature (figure 1). The similar character of the R_H variation with temperature has been observed earlier for indium-doped PbTe when the indium level was resonant with the conduction band [1]. So, one can suppose, that in $Pb_{1-x}Sn_xTe$:Ga alloys the behavior of Hall coefficient also associated with existence of gallium-induced resonant level E_{Ga}, stabilizing the Fermi level on the background of the conduction band, and indicates flowing of electrons from the conduction band to the gallium level due to the decrease of the gap with the increase of the temperature.

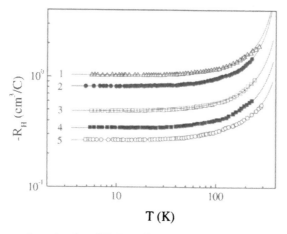

Figure 1. Temperature dependencies of Hall coefficient in n-Pb$_{1-x}$Sn$_x$Te:Ga (x: 1 – 0.09, 2 – 0.14, 3 – 016, 4 – 0.18, 5 – 0.21). Solid lines are calculated according to (1)-(4).

As the tin concentration in the alloys increases the free electron concentration in the samples, determined from the Hall coefficient at helium temperature as n=1/eR$_H$, monotonously increases, indicating the shift of the gallium resonant level with respect to the bottom of conduction band and redistribution of electrons between the level and conduction band. The hydrostatic compression induces the similar behavior of free electron concentration - under pressure the increase of the electron concentration n takes place (figure 2). An additional electron concentration Δn=n(P)-n(0), which is equal to the density of electrons flowing from the level to the conduction band, attains the value higher than $7 \cdot 10^{18}$ cm^{-3}.

DISCUSSION

The obtained experimental results allow us to conclude that doping of Pb$_{1-x}$Sn$_x$Te alloys with gallium leads to appearance of resonant gallium level E$_{Ga}$ on the background of the conduction band. The energy position of resonant level E$_{Ga}$ depends on the temperature, the tin content and the pressure. In order to obtain these dependences and to built the diagrams of resonant impurity level movement relative to the bottom of conduction band we have calculated the dependencies of the electron concentration and the Fermi level position upon temperature, alloy composition and pressure using the temperature dependencies of the Hall coefficient in the frame of two-band Kane's dispersion law for A^{4}B^6 semiconductors [9]:

$$\left(\frac{E_g}{2} - E\right)\left(-\frac{E_g}{2} - E\right) = E_\perp \frac{p_\perp^2}{2m_0} + E_\parallel \frac{p_\parallel^2}{2m_0} \qquad (1)$$

Where E$_g$ is the energy gap, p$_\perp$ and p$_\parallel$ are transversal and longitudinal components of the quasi-

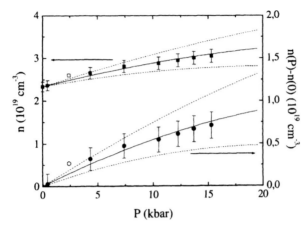

Figure 2. Pressure dependencies of electron concentration n and additional electron concentration $\Delta n = n(P) - n(0)$ in $n\text{-}Pb_{1-x}Sn_xTe\text{:}Ga$ (x=0.21) (open points were obtained under the decrease of pressure). Lines are calculated according to (1)-(4). $d(E_{Ga}-E_i)/dP$, meV/kbar: 1.0 – solid lines, 1.0±0.5 – dotted lines.

impulse relative to the <111> axis, respectively, m_0 is a free electron mass, $E_\perp$ and $E_\parallel$ are the model parameters, characterizing an interaction between the valence and conduction bands (for $Pb_{1-x}Sn_xTe$ (x≈0.2) they are found to be equal $E_\perp \approx 7.65$ eV and $E_\parallel \approx 0.73$ eV [10]).

The temperature, composition and pressure dependencies of the gap were calculated as [9]:

$$E_g(meV) = 171.5 - 535x - 7.4P(kbar) + \sqrt{12.8^2 + 0.19(T(K) + 20)^2} \tag{2}$$

The concentration of free electrons n in the case of degenerate statistics was determined taking into account, that Fermi surface in the p-space consists of the four ellipsoids centered at the L points of the Brillouin zone:

$$n = \frac{1}{eR_H} = \frac{2 \cdot 4}{(2\pi\hbar)^3} \cdot \frac{4\pi}{3} p_\perp^2(E_F) \cdot p_\parallel(E_F) \tag{3}$$

Where $\hbar$ is a Planck's constant.

Using the n value, the Fermi energy relative to the middle of the gap E_i was calculated as:

$$E_F = \sqrt{\left(\frac{3}{4} \cdot \frac{n\pi^2\hbar^3 E_\perp \sqrt{E_\parallel}}{(2m_0)^{3/2}}\right)^{2/3} + \frac{E_g^2}{4}} \tag{4}$$

It was found that theoretical dependencies obtained in such way (lines in figures 1, 2) are in good agreement with experimental data over the entire investigated ranges of temperature, matrix composition and pressure. Assuming a stabilization of the Fermi level by the resonant gallium level E_{Ga} in the conduction band for alloy composition range x=0.09-0.21 we have built the diagram of impurity level movement relative to the principal band edges (figure 3). With increasing tin content in the alloy resonant level moves relative to the conduction band bottom L_6^- with the rate approximately 5 meV/mol.%, remaining almost fixed relative to the valence band top. The temperature rate of gallium level movement relative to the middle of the gap E_i depends on the matrix composition: $d(E_{Ga}-E_i)/dT \approx 0.16-1.13x$ meV/K.

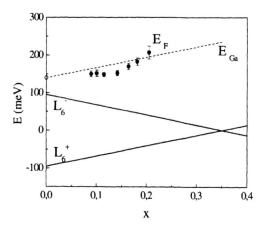

Figure 3. Energy level diagram for $Pb_{1-x}Sn_xTe$:Ga alloys under variation of matrix composition (open point – extrapolation of the data for $Pb_{1-x}Sn_xTe$:Ga alloys [6, 7]).

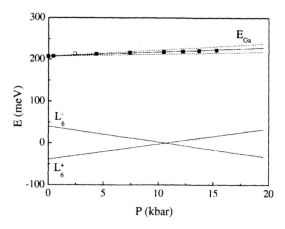

Figure 4. Energy level diagram for $Pb_{1-x}Sn_xTe$:Ga (x=0.21) under pressure (open point was obtained under the decrease of pressure). $d(E_{Ga}-E_i)/dP$, meV/kbar: 1.0 – solid line, 1.0±0.5 – dotted lines.

Pressure dependence of Fermi level position E_F allow us to obtain the diagram of energy spectrum reconstruction for $Pb_{1-x}Sn_xTe:Ga$ (x=0.21) under hydrostatic compression (figure 4). It is shown that under pressure resonant level moves almost linearly slightly shifting upward relative to the middle of the gap: $d(E_{Ga}-E_i)/dP=1.0\pm0.5$ meV/kbar. The similar behavior of Fermi level position under pressure has been observed earlier for $Pb_{1-x}Sn_xTe:In$ and $PbTe:Cr$ [1, 2]. So we suppose that it is an unambiguous proof of an existence of gallium-induced resonant level E_{Ga}, stabilizing the Fermi level on the background of the conduction band of investigated alloys.

CONCLUSIONS

In the present work gallium-induced resonant level situated on the background of conduction band in $Pb_{1-x}Sn_xTe$ alloys was revealed. The galvanomagnetic properties of investigated alloys were explained taking into account stabilization of the Fermi energy by the gallium resonant level, redistribution of electrons between the resonant level and conduction band with the increase of the temperature and movement of the resonant gallium level relative to the conduction band bottom with the change of the tin concentration and under pressure. The temperature, composition and pressure dependencies of electron concentration and Fermi energy in the alloys were obtained and the diagrams of the energy spectrum reconstruction under variation of matrix composition and pressure were proposed.

ACKNOWLEDGMENTS

This research was carried out under financial support from the Russian Foundation for Basic Research (Grant No 05-02-17119).

REFERENCES

1. V.I. Kaidanov and Yu.I. Ravich, Sov. Phys. Usp. **28**, 31 (1985).
2. B.A. Volkov, L.I. Ryabova, D.R. Khokhlov, Physics-Uspekhi **45**, 819 (2002).
3. F.F. Sizov, S.V. Plyacko, V.M. Lakeenkov, Sov. Phys. Semicond. **19**, 368 (1985).
4. G.S. Bushmarina, B.F. Gruzinov, T.T. Dedegkaev, I.A. Drabkin, T.B. Zhukova, and E.Ya. Lev, Inorg. Mater. **16**, 1460 (1980).
5. Z. Feit, D. Eger, A. Zemel, Phys. Rev. **B31**, 3903 (1985).
6. E.P. Skipetrov, E.A. Zvereva, O.S. Volkova, E.I. Slyn'ko, A.M. Mousalitin, Mater. Sci. Eng. **B91-92**, 416 (2002).
7. E. Skipetrov, E. Zvereva, B. Kovalev, E. Slyn'ko, Phys. Stat. Sol. (b) **241**, 120 (2004).
8. V.E. Slyn'ko, Visnyk Lviv Univ., Ser. Physic. **34**, 291 (2001).
9. R. Dornhaus, G. Nimtz and B. Schlicht, *Narrow-Gap Semiconductors* (Springer-Verlag, Berlin, 1983).
10. B.A. Akimov, R.S. Vadhva and S.M. Chudinov, Sov. Phys. Semicond. **12**, 1146 (1978).

Synthesis and Growth

Mater. Res. Soc. Symp. Proc. Vol. 929 © 2006 Materials Research Society   0929-II02-01

Evidence for a Structural Transition to a Superprotonic CsH$_2$PO$_4$ Phase Under High Pressure

Cristian E. Botez[1], Russell R. Chianelli[2], Jianzhong Zhang[3], Jiang Qian[3], Yusheng Zhao[3], Juraj Majzlan[4], and Cristian Pantea[5]

[1]Department of Physics, University of Texas at El Paso, El Paso, Texas, 79968

[2]Department of Chemistry and Materials Research Institute, University of Texas at El Paso, El Paso, Texas, 79968

[3]Los Alamos Neutron Scattering Center, Los Alamos National Laboratory, Los Alamos, New Mexico, 87545

[4]Institute of Mineralogy and Geochemistry, University of Freiburg, Freiburg, 79104, Germany

[5]Materials Science and Technology - National High Magnetic Field Laboratory, Los Alamos National Laboratory, Los Alamos, New Mexico, 87545

ABSTRACT

We have used synchrotron X-ray powder diffraction (SXRPD) to investigate the structural behavior of CsH$_2$PO$_4$ upon heating. Temperature-resolved data collected at *ambient-pressure* demonstrate that a transition from the room-temperature monoclinic phase (*P2$_1$/m*; a=7.90Å, b=6.39Å, c=4.87Å, and β=107.64°) to a high-temperature cubic phase (*Pm3m*; a=4.96Å) occurs at T=237°C. The high-temperature phase is not stable under ambient-pressure conditions, even in the absence of further heating. On the other hand, SXRPD measurements carried out under *high-pressure* (~1GPa) evidence a transition from monoclinic to a *stable* cubic phase (*Pm3m*, a=4.88Å) at a temperature within the 255°C-275°C range. A 1000-fold increase in the proton conductivity (indicating the transition to the superprotonic phase) was previously observed under *the same* non-ambient conditions [1]. Therefore, our results represent strong evidence that the superprotonic behavior in CsH$_2$PO$_4$ is associated with a monoclinic-to-cubic polymorphic structural transition and not with chemical modifications.

INTRODUCTION

In the quest for alternative sources of energy fuel cell research plays a prominent role, as fuel cells hold great potential for a highly efficient and environmentally friendly way of producing electrical energy. The functioning of any fuel cell is based on the electrolyte's ability to conduct ions between the electrodes, but not allow electrons to pass through. It has been recently shown that MH$_n$XO$_4$-type solid acids (M – monovalent cation, n=1,2 and X=S,P) can function as fuel cell electrolytes *at temperatures between 150 and 300°C* [2,3]. Since polymer electrolyte fuel cells cannot operate within this temperature range, solid acids seem to be promising candidates for fuel cell applications where cell functioning at intermediate temperatures is desired (e.g. in the automobile industry). To function as fuel cell electrolytes, solid acids need to undergo a so-called superprotonic phase transition, where, upon heating, their proton conductivity sharply increases by up to four orders of magnitude.

CsHSO$_4$ was the first solid acid used as a fuel cell electrolyte [2], but it was soon noticed that phosphate-based compounds (such as CsH$_2$PO$_4$ and RbH$_2$PO$_4$) would yield better and longer-term cell stability, in part by excluding the possibility of H$_2$S catalyst poison formation [3]. For these phosphate-based (fully-hydrogen-bonded) solid acids, however, the microscopic aspects of the superprotonic behavior are much less understood than for their sulfate-based counterparts. In fact, the mere nature of the superprotonic phase of CsH$_2$PO$_4$ is still under debate. Some authors indicate a polymorphic phase transition [1,4-6], while others associate the superprotonic behavior with chemical changes such as water loss and polymerization [7,8]. Here we present results from synchrotron X-ray powder diffraction experiments aimed at getting more insight into the structural aspects of the superprotonic behavior of this material.

According to previous work [1,5,6], the main experimental challenge in the study of the high-temperature behavior of CsH$_2$PO$_4$ is to prevent its chemical decomposition, which, under ambient pressure and humidity, occurs at 227°C (slightly below the superprotonic transition temperature of 235°C). It has been suggested that decomposition can be avoided by either 1) keeping the sample under a H$_2$O-saturated atmosphere, or 2) subjecting the sample to a pressure of 1.0±0.2 GPa. Using synchrotron X-ray powder diffraction, we have monitored the evolution of the crystal structure of CsH$_2$PO$_4$ upon heating under *both* ambient- *and* high-pressure conditions. We found a monoclinic-to-cubic structural phase transition at T=237°C, even when the heating was carried out under ambient pressure. Yet, the cubic phase decomposed relatively quickly under these conditions. When the temperature was raised while subjecting the sample to high-pressure, however, we observed that the same monoclinic-to-cubic transition led to a high temperature phase that was *stable* not only at the transition temperature but also upon further heating up to 300°C. Our high-pressure data showed that the high-temperature cubic phase of CsH$_2$PO$_4$ (*Pm3m*, a=4.88Å) appeared upon heating under a pressure of ~1GPa in the 255°C-275°C temperature range, *the same* non-ambient conditions under which an abrupt increase in the proton conductivity (indicating the transition to the superprotonic phase) had previously been observed [1]. Consequently, our results represent direct evidence that the superprotonic behavior in CsH$_2$PO$_4$ is associated with a polymorphic structural transition and is not due to chemical modifications.

EXPERIMENTAL PROCEDURE

CsH$_2$PO$_4$ monocrystals were grown by slow evaporation from an aqueous solution prepared by mixing stoichiometric amounts of H$_3$PO$_4$ and Cs$_2$CO$_3$. Crystals were subsequently ground to a fine powder. Ambient-pressure SXRPD measurements were performed on the X7B beamline at the National Synchrotron Light Source (NSLS), Brookhaven National Laboratory using X-rays of wavelength 0.922Å selected by a double flat-crystal monochromator. A Mar345 flat image plate was used to detect the diffracted beam, and diffraction images were collected upon heating in steps of 5°C from room temperature to 250°C. An exposure time of 45 s was used at each temperature. Eventually, the images were processed by integrating over the projections of the Debye-Scherrer cones on the flat detector, using the Plot2D software [9]. High-pressure SXRPD experiments were carried out on the X17B2 beamline at the NSLS. Samples were contained in a DIA-type, large-volume apparatus (SAM85) designed for in-situ X-ray diffraction studies at high pressures and high temperatures.

Samples were initially compressed at room temperature to 1.02 GPa, followed by heating in steps of 20-30 °C up to the a temperature of 350°C. At each temperature step, powder diffraction data corresponding to a 1-6Å d-spacing range were collected. An energy-dispersive X-ray method was employed, where white radiation from a superconducting wiggler magnet was collimated to dimensions of 100 ×100 μm and directed onto the sample. The diffracted X-rays were detected at a fixed angle of $2\theta = 6.4856°$, and the counting time for the collection of each diffraction pattern was 5 minutes.

RESULTS AND DISCUSSION

Figure 1.a shows the room-temperature SXRPD pattern of CsH_2PO_4 measured under ambient-pressure conditions. All peaks were indexed using the unit cell and space group corresponding to the known crystal structure of the room-temperature phase of the title material ($P2_1/m$; a=7.90Å, b=6.39Å, c=4.87Å, and β=107.64°) [10]. We found that this phase persisted during heating up to 232°C, with no unindexed peaks at any temperature in the 25°C-232°C range. At 237°C, the SXRPD pattern changed dramatically (Figure 1.b), suggesting a structural transition to a high-symmetry phase. Indeed, all the peaks at 237°C correspond to a cubic structure ($Pm3m$; a=4.96Å).

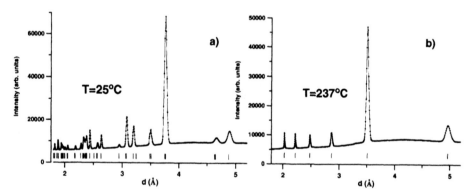

Figure 1. Synchrotron X-ray powder diffraction patterns of CsH_2PO_4 collected under ambient-pressure conditions show a transition from the room-temperature monoclinic phase (a) to a high-temperature cubic phase (b). The transition occurs at T=237°C. The vertical bars indicate the d-spacing positions of the Bragg reflections.

The existence of a cubic high-temperature phase of CsH_2PO_4 was first proposed in an optical microscopy study by Baranov *et al.* [4]. More recent laboratory X-ray diffraction studies by Preisinger *et al.* [5] showed a monoclinic-to-cubic transition at about 230°C for samples heated under a H_2O-saturated atmosphere. However, the same study indicated that, under ambient conditions, no transition was observed due to the dehydration and chemical decomposition of the sample that initiated below the transition temperature. Our experiment demonstrates for the first time that a cubic high-temperature single phase of CsH_2PO_4 could be observed even in the absence of H_2O-saturated atmosphere conditions. However, this phase turned out not to be stable in time, as described below.

Figure 2 shows time-resolved SXRPD patterns of CsH_2PO_4 measured under ambient pressure and humidity at 237°C. These data demonstrate that the high-temperature cubic phase of CsH_2PO_4 is unstable, most likely due to chemical modifications and dehydration, whose effects on the SXRPD patterns are evident after less than 15 min. from the time when the transition occurred. This relatively short time may explain the differences between our results and those of Preisinger *et al.* [5] , as their long data collection time (probably of the order of several hours for each temperature) may have prevented the observation of the short-lived cubic phase under ambient conditions. By using a technique which allows rapid data collection we demonstrated that 1) a complete transition from the room-temperature monoclinic to a high-temperature cubic phase of CsH_2PO_4 occurs even in the absence of humid conditions and 2) the transition *precedes* the onset of the dehydration/chemical decomposition of the title compound.

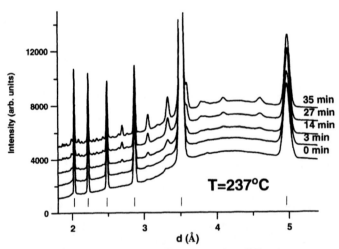

Figure 2. Time-resolved synchrotron X-ray powder diffraction patterns of CsH_2PO_4 demonstrate that, under ambient-pressure conditions, the high-temperature cubic phase is not stable at T=237°C even in the absence of further heating.

To clarify the superprotonic behavior of CsH_2PO_4, Boysen et al. carried out proton conductivity measurements under high-pressure ($\sim$1GPa), a method that allowed them to suppress dehydration at high-temperatures. They observed a three-order-of-magnitude jump in the proton conductivity upon heating to 260°C. In Figure 3, we present our SXRPD results obtained upon heating under high-pressure conditions similar to the ones employed in the above-mentioned proton conductivity study. The gray line represents the powder diffraction pattern collected at T=204°C, whereas the black line represents data collected at T=275°C. The two data sets contain several common peaks, which are fluorescence lines from cesium (marked by crosses) and tungsten (marked by circles) - we see tungsten because it was used in the optics of the multi-element detector. The pattern measured at 204°C corresponds to the room-temperature monoclinic phase of CsH_2PO_4. As the temperature is raised, this phase disappears a within the 255°C-275°C range. Concomitantly, a new SXRPD pattern (with its strongest peak at about 3.4Å), emerges. This high-temperature SXRPD pattern can be indexed to a cubic unit cell (a=4.88Å) and space group $Pm3m$. Thus, our data demonstrate that a monoclinc cubic structural transition occurs under the same pressure and temperature conditions where the onset of the superprotonic behavior had been observed. This represents strong evidence that the dramatic enhancement of the proton conductivity of CsH_2PO_4 at high-temperatures is due to a structural phase transition and not to dehydration and chemical modifications. In fact, dehydration is suppressed by the application of high-pressure, fact confirmed by our observation that, under high-pressure conditions, the cubic phase is stable at 275°C, and upon further heating to 300°C.

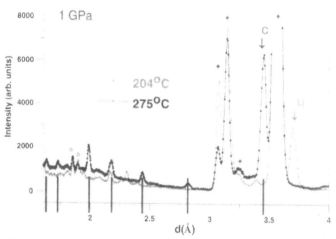

Figure 3. High-pressure synchrotron X-ray powder diffraction patterns of CsH_2PO_4 show a monoclinic-to-cubic structural phase transition that occurs in the 255C°-275C° temperature range. The vertical bars indicate the d-spacing position of the Bragg reflections from the ($Pm3m$; a=4.88Å) cubic phase. The peaks marked by $^+$ and $^°$ are fluorescence lines from cesium and tungsten respectively.

SUMMARY

We have investigated the evolution of the crystal structure of the superprotonic conductor CsH_2PO_4 upon heating under ambient- and high-pressure conditions. Our temperature-resolved synchrotron X-ray powder diffraction data indicate a phase transition from the room-temperature monoclinic phase to a high-temperature cubic phase. When the heating is carried out under high-pressure (~1GPa) the cubic phase appears in the 255°C-275°C temperature range and is stable up to 300°C. Our findings, corroborated with previous results from proton conductivity [1], optical microscopy [4], and laboratory X-ray diffraction [5,6], demonstrate that the superprotonic behavior in CsH_2PO_4 is due to a polymorphic structural transition and not to water loss and chemical modifications.

ACKNOWLEDGEMENTS

CEB would like to acknowledge the support of the University of Texas Research Fund. Use of the National Synchrotron Light Source, Brookhaven National Laboratory, was supported by the U.S. Department of Energy, Office of Basic Energy Sciences, under Contract No. DE-AC02-98CH10886. The "Gateway" Program supported by the U.S. Department of Energy is also acknowledged.

REFERENCES

1. D.A. Boysen, S.M. Haile, H. Liu, and R.A. Secco, Chem. Mater. **15**, 727(2003).
2. S.M. Haile, D.A. Boysen, , C.R.I. Chisholm, and R.B. Merle, Nature **410**, 910(2001).
3. D.A. Boysen, T. Uda, C.R.I. Chisholm, and S.M. Haile, Science, **303**, 68(2004).
4. A.I. Baranov, V.P. Khiznichenko, and L.A. Shuvalov, Ferroelectrics **100**, 135(1989).
5. A. Preisinger, K. Mereiter, and W. Bronowska, Mat. Sci. Forum **166-169**, 511(1994).
6. W. Bronowska, J. Chem. Phys. **114**, 611(2001).
7. J.-H. Park, Phys. Rev. B **69**, 054104(2004).
8. E. Ortiz, R.A. Vargas, and B.-E. Mellander, J. Chem. Phys. **110**, 4847(1999).
9. A.P. Hammersley, S.O. Svenson, M. Hanfland, A.N. Fitch, and D. Hauserman, High Pressure Research **14**, 235(1996).
10. H. Matsunaga, K. Itoh, and E. Nakamura, J. Phys. Soc. (Japan) **48**, 2011(1980).

Mater. Res. Soc. Symp. Proc. Vol. 929 © 2006 Materials Research Society   0929-II02-02

Time Resolved Dynamics of Femtosecond Laser Ablation of Si (100) with Thin Thermal Oxide Layers (20 - 1200 nm)

Joel P. McDonald[1], Vanita R. Mistry[2], John A. Nees[3], and Steven M. Yalisove[4]

[1]Applied Physics, University of Michigan, Gerstacker Building, RM B122, 2200 Bonisteel Ave., Ann Arbor, MI, 48109-2099

[2]Mechanical Engineering, University of Michigan, 2250 G.G. Brown, 2350 Hayward, Ann Arbor, MI, 49109-2125

[3]Center for Ultrafast Optical Science, University of Michigan, 1006 Gerstacker Building, Ann Arbor, MI, 48109

[4]Materials Science and Engineering, University of Michigan, H. H. Dow Building, 2300 Hayward St., Ann Arbor, MI, 48109-2136

ABSTRACT

Femtosecond (fs = 10^{-15} sec.) laser ablation of Si(100) with thermally grown oxide films was studied with pump/probe imaging techniques in order to determine the role of film thickness on ablation dynamics. Two different imaging geometries were used in this study. *Front* view images were formed with the reflection of a delayed probe pulse from the area of a sample irradiated with a pump pulse. By changing the delay between the pump and probe pulses, images were obtained showing the evolution of the surface as a function of time (0 – 12 ns after the arrival of the pump pulse). The *side* view imaging technique, also known as shadowographic imaging, an image was formed of a delayed probe pulse which passed through the ablation plume produced by a pump pulse parallel to the sample surface. Both laser induced shock wave propagation and material removal were observed to change with increased thermal oxide thickness.

INTRODUCTION

The dynamics of ultrafast laser ablation of a variety of materials has been studied with pump-probe imaging techniques. The first such study characterizing femtosecond (fs) laser damage of Si was performed by Downer *et al* in 1985 [1]. Following the initial work in 1985, time resolved ablation dynamics of metals (Ti, Al, Au) and GaAs were conducted with the ultrafast pump-probe imaging technique [2,3]. Time resolved *side* view or shadowographic imaging of ablation of Si and Cu by picosecond (ps) and nanosecond (ns) laser pulses was performed by Russo *et al* in 1999 [4], followed by an investigation using fs laser pulses by Choi *et al* in 2002 [5]. From the combined results of these works, much has been learned about the interaction of intense, ultrafast laser pulses with materials, including the physical mechanisms associated with fs laser induced ablation [3], as well as plasma densities and temperatures on a picosecond time scale [4,5].

In this work, the time resolved ablation dynamics of Si(100) with thermally grown oxide films ranging in thickness from 20 nm to 1200 nm was investigated using pump-probe imaging techniques. The fs laser induced damage morphology of Si(100) with thermal oxide films has been characterized previously with ex-situ optical and atomic force microscopy [6,7], demonstrating that the interaction of the fs laser pulse with the material results in either the removal of the film producing a damage crater of depth approximately equal to the thickness of the film, or blister formation in which the thermal oxide film remains intact but is delaminated from the Si(100) substrate producing a bubble like feature. Pump-probe imaging techniques using 150 fs laser pulses provided an opportunity to study the ablation mechanisms responsible for the observed damage morphologies. In this work, characteristics of the ablation dynamics are discussed with respect to the thickness of the thermal oxide films

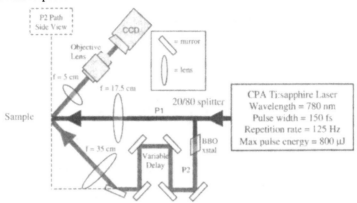

Figure 1. Experimental setup. Laser pulse is incident from right (775 nm, 150 fs pulse width). A 20/80 beam splitter directs 20% of the incident beam to the pump line P1, while the remaining 80% goes to the probe line P2. A BBO crystal is used to frequency double P2 (so that $\lambda = 390$ nm), and the relative delay between the arrival of P1 and P2 at the sample is controlled with the variable delay (mirrors mounted on manual translation rail). The probe path for side view imaging is represented by the dashed line.

EXPERIMENTAL DETAILS

An amplified Ti:sapphire laser was used for all experiments (CPA-2001, Clark-MXR). This laser produced pulses with a maximum energy of 800 μJ, temporal width of 150 fs, and a central wavelength of 775 nm. The experimental setup is presented in figure 1, where it is shown that the beam exiting the laser was subsequently split into two optical paths directed toward the sample, namely the pump (P1, $\lambda = 775$ nm) which produced the ablation event, and probe (P2, $\lambda = 390$ nm) which was used to make an image of the ablation event. For *front* view imaging, P2 was directed onto the sample surface at 45° relative to the sample normal, and an image was formed with the reflection of P2 from the area sample surface irradiated with P1. For *side* view imaging, P2 was directed parallel to the sample surface (90° to sample normal) and a shadowgraph of the ablation event was produced. The image exposure time in each case was dictated by the 150 fs temporal pulse width of P2. A four pass optical delay line was used to delay the arrival of probe pulse with respect to the pump, allowing for images to be collected up to 12 ns after the ablation was initiated. The laser fluence of P1 was 1-3 J/cm^2, while the laser fluence of P2 was adjusted for optimal imaging conditions, typically around 0.1 J/cm^2.

RESULTS

Front view imaging

Front view images were collected from Si(100) samples with 20 nm and 1200 nm of thermal oxide using a pump fluence of 2.5 J/cm^2 and 2.1 J/cm^2 respectively. A series of images collected of the ablation event for each case is presented in figure 2. The rings of concentric contrast evident in the images are known as Newton's rings, and are an indication of the material expansion resulting from the ablation event [2,3]. The images presented in figure 2 for each oxide thickness show the emergence of three such rings as a function of time, with the beginning

a) 20 nm thermal oxide, pump fluence of 2.5 J/cm².

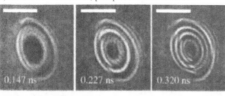

b) 1200 nm thermal oxide, pump fluence of 2.1 J/cm².

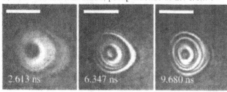

Figure 2. Front-view images of fs laser induced ablation of Si(100) with thermal oxide at time delays shown. a) 20 nm thermal oxide, pump fluence of 2.5 J/cm². b) 1200 nm thermal oxide, pump fluence of 2.1 J/cm². Scale bar in each image is 50 μm.

of a new ring indicated by an interference minimum or dark spot in the center of the feature. The temporal evolution of the rings was found to be highly dependent on the oxide thickness (i.e the first interference minimum occurred at 0.147 ns for the 20 nm film, while the first minimum appeared at 2.613 ns for the 1200 nm film), suggesting that the expansion resulting from the ablation event was slowed by the presence of thicker thermal oxide film. A calculation of the material expansion rate at the center of the ablation event based on the temporal evolution of the Newton's rings provided a range for the expansion rate of the ablated material. A precise value for the expansion rate was not obtained however, as analysis required knowledge of the optical

Edge Diffraction

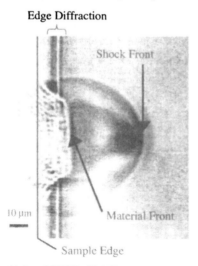

Figure 3. Side-view image of fs laser ablation of Si(100) with 147 nm thermal oxide film at a laser fluence of 3.0 J/cm², collected 7.01 ns following absorption of the pump pulse at the sample surface.

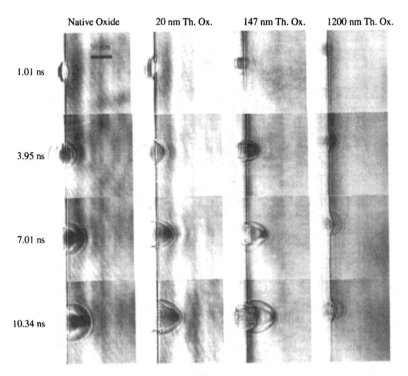

Figure 4. Side-view images of fs laser ablation of Si(100) with oxide films of varying thickness (indicated at top) at a laser fluence of 3.0 J/cm^2, collected over a range of times (indicated at left) following absorption of the pump pulse at the sample surface. The scale bar shown in the upper left applies to all images, and note that all images have been inverted with respect to the originals.

properties of the materials during ablation, which were unknown.

Side view imaging

Side view images of fs laser ablation of Si(100) were collected from samples with native oxide, as well as 20 nm, 147 nm, and 1200 nm of thermally grown oxide using a laser fluence of 3.0 J/cm^2. These images showed the presence of two primary fronts, illustrated in figure 3. Please note that all side view images presented are inverted with respect to the originals. Examples of these images for the range of oxide thickness and time delay are presented in figure 4. Also present in these images are vertical fringes due to diffraction as the probe pulse passes the edge of the sample [8], and an imaging artifact present interior to the sample edge which is not fully understood. The shock front is generated by the sudden absorption of laser energy which produces a sudden thermal expansion of the material surface [9]. The material front is the leading edge of the material removed or ablated by the action of the incident pump pulse. The physical nature of the two fronts was verified using laser induced breakdown spectroscopy in an orthogonal, dual pulse configuration.

Qualitatively, the images in figure 4 demonstrate the influence of the thermal oxide thickness on the ablation dynamics. It can be seen that both the shape and height of the shock front for a particular time delay depend on the thickness of the glass. Additionally, the propagation of the material front appears to depend on the thermal oxide thickness. For example, for the case of the 20 nm thermal oxide film, the material front and shock front remain close together over the range of delays presented, while in all other cases the material front and shock front are evidently separated. A quantitative analysis of the shock expansion dynamics [4,5] will provide additional insight into the differences in the fs laser induced ablation as a function of thermal oxide thickness.

CONCLUSIONS

The presence of thermally grown oxide films (20 nm to 1200 nm) on Si(100) was found to modify fs laser induced ablation by influencing both shock dynamics and material removal. Future work will address analysis of shock propagation as a function of laser fluence and oxide thickness.

ACKNOWLEDGEMENTS

The authors would like to acknowledge support from the Air Force (MURI-AFOSR grant #A9550-05-1-0416) and the National Science Foundation (grant # DMR03070400).

REFERENCES

[1] M. C. Downer, R. L. Fork, C. V. Shank, J. Opt. Soc. Am. B. **2**, 595 (1985).
[2] K. Sokolowski-Tinten, J. Bialkowski, A. Cavalleri, D. von der Linde, A. Oparin, J. Meyer-ter-Vehn, S. I. Anisimov, Phys. Rev. Lett. **81**, 224 (1998).
[3] D. von der Linde, K. Sokolowski-Tinten, Appl. Surf. Sci. **154**, 1 (2000).
[4] R. E. Russo, X. L. Mao, H. C. Liu, J. H. Yoo, S. S. Mao, Appl. Phys. A 69 [Suppl.], S887 (1999).
[5] T. Y. Choi, C. P. Grigoropoulos, J. Appl. Phys. **92**, 4918 (2002).
[6] J. P. McDonald, V. R. Mistry, K. E. Ray, S. M. Yalisove, Mater. Res. Soc. Symp. Proc. **875**, O.12 (2005)
[7] J. P. McDonald, V. R. Mistry, K. E. Ray, N. R. Moody, J. A. Nees, S. M. Yalisove, Appl. Phys. Lett. **88**, 153121 (2006).
[8] M. Born, E. Wolf, in *Principles of Optics* (Cambridge, United Kingdom, 1999).
[9] T. W. Murray, J. W. Wagner, J. Appl. Phys. **85**, 2031 (1999).

Mater. Res. Soc. Symp. Proc. Vol. 929 © 2006 Materials Research Society   0929-II02-03

Spatially Resolved MicroDiffraction Analysis of the Plastic Deformation in the Shock Recovered Al Single Crystal

R. I. Barabash[1,2], G. E. Ice[1], W. Liu[3], J. Belak[4], and M. Kumar[4]

[1]Materials Science and Technology, Oak Ridge National Laboratory, One Bethel Valley Road, Oak Ridge, TN, 37831

[2]Center for Materials Processing, University of Tennessee, Knoxville, TN, 37996

[3]Advanced Photon Source, Argonne, IL, 60439

[4]University of California, Lawrence Livermore National Laboratory, Livermore, CA, 94551

Abstract

Spatially resolved diffraction with a sub micrometer-diameter beam and 3D differential aperture microscopy together with MD simulations, SEM and OIM analysis are applied to understand the arrangements of voids, geometrically necessary dislocations and strain gradient distribution in samples of Al (123) single crystal shocked to incipient spallation fracture. We describe how geometrically necessary dislocations and the effective strain gradient alter white beam Laue patterns of the shocked materials. We show how to quantitatively determine the orientation and density of geometrically necessary dislocations in the shock recovered Al samples being initially oriented for single slip.

Introduction

Strong shock waves result in the transition from elastic to plastic compression. As a result of dislocation motion and the strong interaction between dislocations and elastic waves, an initially random dislocation distribution becomes unstable and forms a correlated dislocation arrangement with dislocation walls. Some fraction of the dislocations may remain randomly distributed, and the rest form various correlated groupings and more organised disclination arrangements. Regions with geometrically necessary dislocations may form causing local lattice curvature.

Aluminum has been the object of numerous shock experiments. Reshock and release of shock in compressed aluminum was studied previously by Lipkin and Asay[1]. Recently developed in-situ time resolved x-ray diffraction relates in-situ pseudo Kossel line broadening with dislocation density during the shock experiments[2 - 4]. Effect of stress triaxiality on void growth in dynamic fracture of metals was studied with molecular dynamic simulations[5]. We complement these results with a white (polychromatic) X-ray microbeam study of the meso-scale geometrically necessary dislocation arrangement in the (123) Al single crystal after shock. The single crystal Al samples were shocked to incipient spallation fracture on the LLNL light gas gun as was done by Stevens et.al.[6]

Experimental

The x-ray synchrotron measurements were made with a 3D X-ray crystal microscope[7] based on a polychromatic (white) beam with an energy range 6 - 25 KeV. This method allows for true 3D mapping of the crystalline phase, orientation and plastic deformation with unprecedented spatial resolution. A region of interest is identified using an optical microscope and the polychromatic x-ray beam is focused onto the sample. The crystallographic orientation at each position of the sample is precisely determined by an automated indexing program. Because polychromatic microdiffraction produces Laue images from each deformation cell intercepted by the x-ray beam with lattice rotations, the typical problem is an overlap of many Laue patterns that must be disentangled. The process of disentangling can be carried out either by a direct experimental technique called differential aperture microscopy or by pattern matching/ absorption sensitive methods[7]. Once the Laue pattern from a subgrain volume has been obtained and the indices of each reflection determined, powerful single-crystal methods can be employed to determine the unit cell parameters. Although differential aperture microscopy (DAXM) can often resolve Laue image into single crystal-like patterns, when the density of unpaired or "geometrically necessary" dislocations (GNDs) is high, streaking is observed even for microsize regions in DAXM resolved patterns[8 - 10]. Typically streaking in Laue patterns indicates the presence of lattice rotations within the probed region as a result of the GNDs formation.

Data collection has been carried out using microbeam Laue diffraction on beamline 34ID at the Advanced Photon Source. Dimensions of the beam were 0.5 by 0.5 microns with a penetration depth of ~ 500 micron. Details on the experimental setting and data collection can be found elsewhere[7 - 10].

A sketch of the shocked Al single crystal and the geometry of the incident and diffracted white beam radiation and the microbeam geometry are shown in the Fig. 1. An (123) Al single crystal with an initial diameter of 5 mm and thickness of 5 mm was cut along the shock direction and analyzed at different locations of the sample cross section (Fig 1b). The cutting was done with a diamond saw and the sample was then chemically etched to remove about 100 micron from the surface and any damage that the diamond saw may have caused.

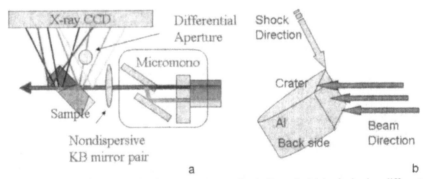

Figure 1. 3D polychromatic microbeam geometry: wire (yellow circle) is shadowing different portions of the diffracted intensity on a CCD (a); sketch of the Al single crystals orientation during shock experiment and orientation relative to microbeam (b).

Results and Discussion

A number of experimental studies of plastic deformation under different loading conditions show that typically a cell-wall structure with alternating regions of high and low dislocations density is formed[9 - 14]. The dislocation walls contain a high density of dislocations, and cell interiors have very low dislocation population. Shock often causes the formation of numerous voids in materials[6].

An approach to characterize the dislocation structure and determine the activated slip systems by analyzing the streaked Laue pattern using a multiscale hierarchical framework is described in detail in the Refs. [8 - 10]. In white-beam diffraction the full-width-at-half-maximum of the streak (ξ direction) $FWHM_\xi \equiv \delta m_\xi$ is a function of the misorientation within the probed region caused by geometrically necessary dislocations. It depends on the orientation and the number of unpaired or geometrically necessary dislocations (GNDs) and/or number of geometrically necessary boundaries (GNBs) in the probed volume. The orientation of the GNDs and/or GNBs influences the character of Laue spot. In the transverse direction v, the $FWHM_v \equiv \delta m_v$ depends on the total number of all dislocations (GNDs and statistically stored) per unit length and usually $\delta m_\xi \gg \delta m_v$.

Under multiple slip the Laue intensity distribution depends on the dislocation density tensor ρ_{ij}. The elements of the dislocation density tensor can be written as:

$$\rho_{ik} = \tau_i b_k \delta(r).$$ (1)

The magnitude of the vector, τ, is the net number of dislocations having Burgers vector, b, crossing unit area normal to τ [15]. Following the approach[16] in the framework of strain gradient plasticity, the total GNDs and GNBs density tensor ρ_{ij} relates to the strain gradient tensor η_{lmk};

$$\varepsilon_{ilm}\eta_{lmk} = -\rho_{ik}; \quad \eta_{lmk} = \frac{\partial^2 u_k}{\partial x_l \partial x_m}$$ (2)

Here ε_{ilm} is the anti-symmetric Levi-Civita tensor. Under multiple slip, the intensity distribution around each reciprocal lattice point and the streak direction ξ of the Laue spot depends on the aforementioned dislocation density tensor. Here u_k, the k-th component of the displacement field u for any unit cell, is due to all dislocations in the crystal.

We calculate the total displacement of the i-th cell u_i from the equilibrium positions R_i^0 corresponding to the undeformed crystal using continuum elastic theory. This displacement is due to all dislocations and is defined using random numbers c_t by the equation[7-9]:

$$\mathbf{u}_i = \sum_t c_t \mathbf{u}_{it}; \quad c_t = \begin{cases} 1,(+) \\ 0,(-) \end{cases}$$ (3)

If there is a dislocation at position t, $c_t = 1$, and for positions without dislocations $c_t = 0$. The dislocation density tensor ρ_{ik} gives the sum of the Burgers vectors of the dislocations whose Burgers vector and line direction are directed parallel to x_k and x_i axes (k, i =1, 2, 3), respectively. These are set as an initial input parameter for simulation of the Laue pattern. By simulating Laue images with different GND slip systems, we can determine the set of slip systems that are best fit to the local lattice curvature and

65

experimentally determined Laue image. The analysis of the dislocation substructure for multiple slip deformation includes the following steps:

- Determination of the local lattice curvature (misorientation axes and angle) from the experimental Laue pattern at each probed location.
- Simulation of Laue pattern using these parameters to check its identity with the original experimental image.
- Performing least square fit of the streak profile find the components of the dislocation density tensor.

The polychromatic x-ray microdiffraction (PXM) technique was applied to a complicated dislocation structure arising from the shock induced plastic/elastic deformation of the Al (123) single crystal. The microbeam-Laue diffraction reveals several distinct zones located at different depth under the shock front (indicated by several red arrows in Fig.1 b). Pronounced streaking of the Laue images is observed in the zones 1 and 6 close to the front and back surface of the shocked (123) Al single crystal (Fig. 2).

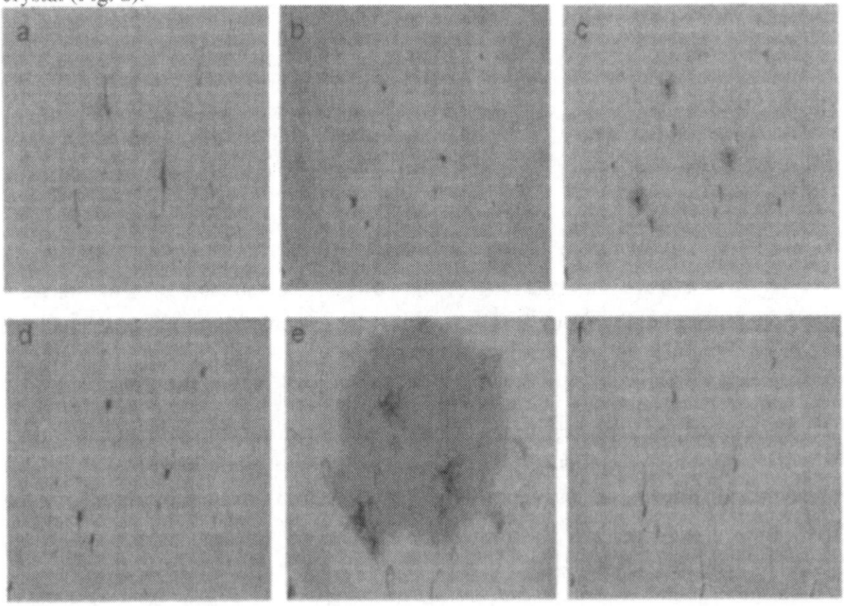

Figure 2. Laue images obtained at different depth under the shock front of (123) Al single crystal: a) zone 1 near the shock front; b).zone 2;c) zone 3; d)zone 4; e)zone 5 with corresponding to the region of intensive void formation; f) zone 6 near the back surface of the shocked single crystal.

Much less streaking is observed in the zone 2, 3, 4 located approximately in the upper half of the sample. The most interesting images correspond the zone 5 (Fig. 2 e). This zone corresponds to the region in which X-ray tomography, SEM and OIM indicated intensive void formation (Fig.3).

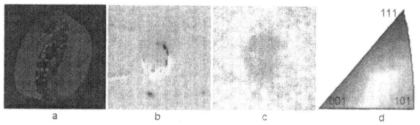

a	b	c	d

Figure 3. X-ray tomography shows region with high concentration of mesoscale size voids in the spall region (a), SEM image of the void at the surface of the cross section of the after shock Al single crystal chosen for PXM analysis, OIM image of the orientation change around the same void (c) and color scale of orientation distributions (d).

To get a better understanding of the reasons for such a peculiar shape of the Laue image, 3D depth resolved measurements were performed. These differential aperture measurements allow tracking the change in the intensity at different depth along the beam. They showed that in this region alternating local lattice rotations take place (Fig.4). We argue that such alternating local rotations are caused by the presence of voids creating regions of plastic deformation in the vicinity of the void. This result agrees well with SEM and OIM observations.

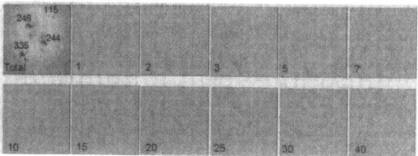

Figure 4. Total (left top corner) and depth resolved Laue patterns from the spall region with voids. Depth is shown in microns at each depth resolved images. Total Laue pattern is indexed for convenience.

Conclusions
Microbeam-Laue diffraction reveals several distinct zones located at different depths under the shock front. Pronounced streaking of Laue images are observed in the zones close to the front and back surface of the shocked (123) Al single crystal. Streaking indicates the formation of geometrically necessary dislocations which is consistent with a single slip mode in the near surface regions. Below the surface of the Al single crystal the portion of geometrically necessary dislocations in the dislocation ensemble reduces, and at the same time the portion of statistically stored dislocations increases. Intensive mesoscale void formation is observed close to the center of the samples, which is

accompanied by a complicated shape of the Laue diffraction images. The density and organization of dislocations as a function of depth under the shock front agrees qualitatively to SEM results on the same samples.

Acknowledgement

The authors have benefited enormously from discussions with Hector Lorenzana, Justin Wark and Jan Ilavsky. Research sponsored by the Division of Materials Sciences and Technology, Office of Basic Energy Sciences, U.S. Department of Energy, under Contract DE-AC05-00OR22725 with UT-Battelle, LLC; at Lawrence Livermore National Laboratory research is sponsored under the Contract W-7405-Eng-48.

References

[1] J. Lipkin and J.R. Asay, *J Appl. Phys* **48** 1, 182-189 (1977).

[2] A. Loveridge-Smith, A. Allen, J. Belak, T. Boehly, A. Hauer, B. Holian, D. Kalantar, G. Kyrala, R.W. Lee, P. Lomdahl, M.A. Meyers, D. Paisley, S. Pollaine, B. Remington, D.C, Swift, S. Weber, and J.S. Wark, *Physical Review Letters*, **86**, 11, 2349-2352 (2001).

[3] D.H. Kalantar, J. Belak, E. Bringa, K. Budil, M. Caturla, J. Colvin, M. Kumar, K.T. Lorenz, R. E. Rudd, and J. Stolken, *Physics of Plasmas*, **10**, 5, 1569-1576 (2003).

[4] D.H. Kalantar, E. Bringa, M. Caturla, J. Colvin, K.T. Lorenz, M. Kumar, and J. Stöölken, *Review Scientific Instruments*, **74**, 3, 1929-1934 (2003).

[5] E.T. Seppälä, J. Belak, and R.E. Rudd, *Physical Review B*, **71**, 64112 (2005); and references therein.

[6] Stevens, A.L., L. Davison and W.E. Warren, J. Appl. Phys. **43**, 4922 (1972).

[7] Larson B.C., Wenge Yang, G.E. Ice, J.D.Budai, J.Z. Tischler, Nature, **415**, 887-890 (2002).

[8] R. Barabash, G.E. Ice, B.C. Larson, G.M. Pharr, K.-S. Chung, W. Yang, Appl.Phys.Lett. **79**, 749 (2001).

[9] R. Barabash, G.E. Ice, F. Walker, J. Appl. Physics, **93**, 3, 1457-1464 (2003).

[10] R.I. Barabash, G.E. Ice, Microdiffraction Analysis of Hierarchical Dislocation Organization. In: Encyclopedia of Materials: Science and Technology Updates, Elsevier, Oxford. 1-18 (2005).

[11] Mughrabi, H., *Acta metall.* **31**, 9, 1367-1379 (1983).

[12] Hansen, N., *Metallurgical and materials transactions A* **32A**, 2917-2935 (2001).

[13] Hughes, D.A and Hansen, N, *Acta Mater.* **48**, 2985-3004 (2000).

[14] Hughes, D.A., Liu, Q., Chrzan, D.C. and Hansen, N., *Acta Mater.* **45**, 105-112

[15] Nye, J.F. (1953) *Acta Metallurgica*, **1**, 153-162 (1997).

[16] Gao, H., Huang, Y., Nix, W. D. and Hutchinson, J. W., *Journal of the Mechanics and Physics of Solids*, **47**, 1239-1263 (1999).

Mater. Res. Soc. Symp. Proc. Vol. 929 © 2006 Materials Research Society   0929-II02-06

Rapid Synthesis of Dielectric Films by Microwave Assisted CVD

Nicholas Ndiege[1], Vaidyanathan Subramanian[2], Mark Shannon[3], and Rich Masel[4]
[1]Chemistry, University of Illinois, 600 S. Mathews ave, Urbana, IL, 61801
[2]Chemical & Biomolecular Engineering, University of Illinois, 600 S. Mathews ave., Urbana, IL, 61801
[3]Mechanical & Industrial Engineering, University of Illinois, Urbana, IL, 61801
[4]Chemical & Biomolecular Engineering, University of Illinois, 600 S. Mathews ave, Urbana, IL, 61801

Typical methods for the deposition of high quality films require high temperatures and/or high vacuum and have characteristically slow deposition rates. In this paper we describe a novel deposition technique: Microwave assisted chemical vapor deposition (MACVD). This technique produces films of high quality at unusually high deposition rates (>1μm a minute) at ambient pressures and temperatures slightly above room temperature. The equipment used is very inexpensive compared to that used in the conventional techniques such as epitaxy, LPCVD, e-beam evaporation, sol-gel. In this study we employ MACVD in the deposition of high k dielectric films. Such materials are typically transition metal oxides with k values such as 65 for α Ta_2O_5 and 14.2 for FeO compared to SiO_2 at ~4.4. [1] The challenges faced in employing these materials as dielectric barriers are mobility degradation and, more significantly, the formation of a silicide interface between the substrate and oxide film that compromises the overall dielectric properties of the film. [2] High k dielectric films (specifically Ta_2O_5) are also appealing in the fabrication of selective emitters for use in thin film insulation and anisotropic heat expulsion for refractory environments. The challenge posed by this application is that fabrication of metal oxide films of over 1μm thickness is plagued by very low growth rates, extensive film cracking, buckling and even catastrophic delamination. [3-5]

Film properties are highly dependent on the method of deposition employed, and up to this time, none of the commonly used techniques (e.g. e-beam assisted deposition, sputtering, sol-gel) has been able to yield film coatings that overcome the limitations thus cited. This has led to the quest for novel and economically viable deposition techniques that will be able to satisfy the demands put forth for microelectromechanical applications, without putting too much strain on the environment. [2, 7, 8]

In this study, we report the deposition of Ta_2O_5 films on silicon via microwave assisted chemical vapor deposition at atmospheric conditions using a domestic microwave oven. The resulting films are also characterized and probed for various chemical, physical and optical properties. Dense and stable films of thickness ranging from 60nm to 62μm have been deposited in this manner. To the best of our knowledge, this is the first time a microwave field has been applied to the substrate and precursor to assist in the ambient metallorganic chemical vapor deposition of any dielectric.

The microwave reactor used in this study was an unmodified 1100W, 2.45GHz domestic oven with two different experiment set-ups, either vertically oriented rectangular (0.6cm x 4cm) or 2cm x 2cm horizontally oriented (fig. 8, fig. 9).

Films deposited on unpolished Si were found to be much thicker than those on polished Si. The crust of the thick films exhibited cracks attributable to the large thermal gradient experienced across the film surface once the microwave field is turned off. The crust is removable by sample polishing (fig. 1(a), (b)). All samples were very stable and highly resistant to scratching by a sharp metal object. The thin films are crack-free, but the thicker films do show some defects though of a very low defect concentration, most likely due to rapid cooling of the sample upon termination of the microwave field. Channels on the patterned substrate were filled with a relatively high degree of uniformity save for the junction of the growth fronts (fig. 1(b). XRD analysis of the as deposited films revealed an amorphous composition while the calcined films showed an orthorhombic phase of Ta_2O_5 (fig. 3). The spectrum matches ICSD card # 97-004-7493.

Depth profile analysis was done on the samples using AES (3 keV ion beam) to examine the elemental composition through the film cross-section and also at the film-substrate interface. The substrates were oriented horizontally during deposition (2nd type of set up) to get a large area of deposition. The uncalcined films were also analyzed for comparison. The 10 sec exposure sample has a thin silicide layer (~29nm) formed upon calcination that is typical of thermally treated transition metal oxide films prepared by conventional techniques (fig.5). [6, 9] There is also a thin Si-O layer (~10nm) underneath the Ta-Si-O layer as expected for Ta_2O_5 films calcined in O_2. [9] The control (uncalcined) film profile was taken for comparison (fig.4) showing a silicide layer formation (~23nm). Similar interface is present in films deposited via MOCVD and reactive sputtering. [6, 9] Film thickness estimates were made from high-resolution SEM images. XPS and AES depth profile analysis of 130 nm films show no carbon present. However, 800 nm films show carbon content of 30% atomic concentration (fig. 10, fig. 11).

SEM cross-section images show a thickness of 150nm before and after calcination (fig. 1 (c)) for the 10sec exposure film. Cross-section images of the 1.5 min exposure film show a thickness of 800nm before calcination that is reduced to 575nm after calcination (28% reduction) (fig. 2). To the best of the authors' knowledge, such thicknesses of dense, stable Ta_2O_5 films have not been achieved within comparable timescales for conventional deposition methods. [3, 10-14]

XPS analysis was done on the calcined films. The surface was sputtered with argon for 12 minutes prior to analysis. No residual carbon was observed (fig. 6). (C 1s signal, if present, appears at 285 eV). Surface profilometry analysis was used to determine thickness profiles of the vertically mounted samples. The probe was dragged across the length of the sample from the tip above the precursor to the base in contact with the precursor. The meniscus of the precursor prior to deposition was at 3.5cm. Deposition is seen to occur on only the bottom half to two thirds of the substrate (fig. 7). This is observed for samples mounted at different heights within the microwave cavity. This result is still under investigation, but the cause is hypothesised to be uneven coupling of the substrate with the microwave field.

There is much debate about the precise effect of the microwave field upon its interaction with matter. [15-17] A microwave photon possesses 0.0016eV, which is not enough energy to break chemical bonds, and is below the energy of Brownian motion. However, a wide range of inorganic materials is known to couple well with microwave fields due to the electric dipoles present in these materials responding to the applied electric field of the microwaves. [18, 19] Thermal/kinetic effects are a result of dipolar polarization and ionic conduction. Oscillation of the applied field causes the matrix's dipole or ion field to attempt realigning itself with the alternating electric field. This process causes energy loss as heat via molecular friction and dielectric loss. No heating occurs if the dipole reorients fast enough with the applied field. Other non-thermal and microwave specific effects are the subjects of much controversy and proof one way or another demands non-trivial experimentation. [20-27]

In conclusion, we have used microwave assisted chemical vapor deposition to generate orthorhombic Ta_2O_5 films on silicon at atmospheric conditions. Sample thicknesses vary from 60 nm to 62μm with unusually high deposition rates. Thin (130nm and below) as deposited films are found to be carbon free. This mode of deposition is a potentially cheap and simple route to fabrication of photonic crystals. It also yields samples with similar properties as samples deposited via conventional albeit expensive techniques. Further investigations are underway to better understand the precise microwave effects on the substrate and precursor. These studies will also elucidate the deposition mechanism underway.

Acknowledgments: AES, SEM, XRD and XPS analysis was carried out in the Center for Microanalysis of Materials, University of Illinois, which is partially supported by the U.S. Department of Energy under grant DEFG02-91-ER45439

This work was supported by the Department of Defense Multidisciplinary University Research Initiative (MURI) program administered by the Army Research Office under contract DAAD19-01-1-0582. Any opinions, findings, conclusions or recommendations expressed in this publication are those of the authors and do not necessarily reflect the views of the Department of Defense or the Army Research Office.

Experimental

An unmodified 1100W, 2.45GHz Emerson® domestic microwave oven was used. Tantalum (v) ethoxide (99.98%) from Aldrich was used as precursor, with deposition on Si <111> substrate. Substrates are degreased in acetone and isopropyl alcohol each for 10 min in ultrasound, followed by a 10min dip in dilute HF and a final H_2O rinse. Two types of set ups were used for mounting the substrate and precursor. In the 1st set up, 1ml of precursor liquid was contained in a 5ml screw-capped vial with the substrate standing upright. The rotating microwave plate was used. Sample vials were placed at the center of the microwave for all runs (fig. 8). The 2nd set up comprised of 0.8ml tantalum (v) ethoxide in a shallow alumina crucible with a Si<111> wafer covering the top of the crucible (~2cm diameter). This setup was mounted with the microwave oven flipped 90° counterclockwise such that the impinging radiation was primarily perpendicular to the substrate and the precursor surface (fig. 9). Deposition on patterned surfaces was also performed using the 1st set up to further investigate the viability of this technique for fabrication of photonic crystals. The substrate was 2cm x 2cm Si <100> die with 1.2cm long, 100μm deep and 100μm wide grooves etched into it.

Repeat runs were conducted at the exact spot to reduce error due to inhomogeneity of the microwave field in the cavity. The 2nd set up allowed for much larger areas of deposition than the 1st. The resulting films were calcined at 900°C for 9hrs under 100cm^3/min flow of oxygen.

Spectroscopic techniques were applied to characterize the films obtained via x-ray diffraction (Rigaku DMAX® XRD with Cu (Kα) radiation), scanning electron microscopy (HITACHI S4700® SEM), x-ray photoelectron spectroscopy (PHI 5400® XPS system with a Mg Kα x-ray source) and auger electron spectroscopy (PHI 660® Auger) analysis. Surface profilometry measurements were also performed on the vertically oriented substrate (Sloan Dektak3 ST®).

References:

1. *High Dielectric Constant Materials: VLSI MOSFET Applications*. ed.; Springer: New York, 2005; 'Vol.' p.
2. Robertson, J., Interfaces and defects of high-K oxides on silicon. *Solid-State Electronics* **2005,** 49, (3), 283-293.
3. Chaneliere, C.; Autran, J. L.; Devine, R. A. B.; Balland, B., Tantalum pentoxide (Ta2O5) thin films for advanced dielectric applications. *Materials Science & Engineering R-Reports* **1998,** 22, (6), 269-322.
4. Joshi, P. C.; Cole, M. W., Influence of postdeposition annealing on the enhanced structural and electrical properties of amorphous and crystalline Ta2O5 thin films for dynamic random access memory applications. *Journal of Applied Physics* **1999,** 86, (2), 871-880.
5. Liu, L.; Wang, Y.; Gong, H., Annealing effects of tantalum films on Si and SiO2/Si substrates in various vacuums. *Journal of Applied Physics* **2001,** 90, (1), 416-420.
6. Atanassova, E.; Konakova, R. V.; Mitin, V. F.; Koprinarova, J.; Lytvym, O. S.; Okhrimenko, O. B.; Schinkarenko, V. V.; Virovska, D., Effect of microwave radiation on the properties of Ta2O5-simicrostructures. *Microelectronics Reliability* **2005,** 45, (1), 123-135.
7. Yoshimura, M., Importance of soft solution processing for advanced inorganic materials. *Journal of Materials Research* **1998,** 13, (4), 796-802.
8. Roberts, B. A.; Strauss, C. R., Toward rapid, "green", predictable microwave-assisted synthesis. *Accounts of Chemical Research* **2005,** 38, (8), 653-661.
9. Ono, H.; Hosokawa, Y.; Ikarashi, T.; Shinoda, K.; Ikarashi, N.; Koyanagi, K.; Yamaguchi, H., Formation mechanism of interfacial Si-oxide layers during postannealing of Ta2O5/Si. *Journal of Applied Physics* **2001,** 89, (2), 995-1002.
10. Ozer, N.; Lampert, C. M., Structural and optical properties of sol-gel deposited proton conducting Ta2O5 films. *Journal of Sol-Gel Science and Technology* **1997,** 8, (1-3), 703-709.
11. Ezhilvalavan, S.; Tseng, T. Y., Preparation and properties of tantalum pentoxide (Ta2O5) thin films for ultra large scale integrated circuits (ULSIs) application - A review. *Journal of Materials Science-Materials in Electronics* **1999,** 10, (1), 9-31.
12. Kukli, K.; Aarik, J.; Aidla, A.; Kohan, O.; Uustare, T.; Sammelselg, V., Properties of Tantalum Oxide Thin-Films Grown by Atomic Layer Deposition. *Thin Solid Films* **1995,** 260, (2), 135-142.

13. Toki, K.; Kusakabe, K.; Odani, T.; Kobuna, S.; Shimizu, Y., Deposition of SiO2 and Ta2O5 films by electron-beam-excited plasma ion plating. *Thin Solid Films* **1996**, 282, (1-2), 401-403.

14. Werder, D. J.; Kola, R. R., Microstructure of Ta2O5 films grown by the anodization of TaNx. *Thin Solid Films* **1998**, 323, (1-2), 6-9.

15. de la Hoz, A.; Diaz-Ortiz, A.; Moreno, A., Microwaves in organic synthesis. Thermal and non-thermal microwave effects. *Chemical Society Reviews* **2005**, 34, (2), 164-178.

16. Kappe, C. O., Controlled microwave heating in modern organic synthesis. *Angewandte Chemie-International Edition* **2004**, 43, (46), 6250-6284.

17. Vigil, E.; Ayllon, J. A.; Peiro, A. M.; Rodriguez-Clemente, R.; Domenech, X.; Peral, J., TiO2 layers grown from flowing precursor solutions using microwave heating. *Langmuir* **2001**, 17, (3), 891-896.

18. *Microwaves in Organic Synthesis*. ed.; Wiley-VCH Verlag GmbH & Co. KGaA: Weinheim, 2002; 'Vol.' p.

19. Katz, J. D., Microwave Sintering of Ceramics. *Annual Review of Materials Science* **1992**, 22, 153-170.

20. Kuhnert, N., Microwave-assisted reactions in organic synthesis - Are there any nonthermal microwave effects? *Angewandte Chemie-International Edition* **2002**, 41, (11), 1863-+.

21. Strauss, C. R., Microwave-assisted reactions in organic synthesis - Are there any nonthermal microwave effects? Response. *Angewandte Chemie-International Edition* **2002**, 41, (19), 3589-3590.

22. Pivonka, D. E.; Empfield, J. R., Real-time in situ Raman analysis of microwave-assisted organic reactions. *Applied Spectroscopy* **2004**, 58, (1), 41-46.

23. Brooks, D. J.; Brydson, R.; Douthwaite, R. E., Microwave-induced-plasma-assisted synthesis of ternary titanate and niobate phases. *Advanced Materials* **2005**, 17, (20), 2474-+.

24. Brooks, D. J.; Douthwaite, R. E., Microwave-induced plasma reactor based on a domestic microwave oven for bulk solid state chemistry. *Review of Scientific Instruments* **2004**, 75, (12), 5277-5279.

25. Getvoldsen, G. S.; Elander, N.; Stone-Elander, S. A., UV monitoring of microwave-heated reactions - A feasibility study. *Chemistry-a European Journal* **2002**, 8, (10), 2255-2260.

26. Stellman, C. M.; Aust, J. F.; Myrick, M. L., Situ Spectroscopic Study of Microwave Polymerization. *Applied Spectroscopy* **1995**, 49, (3), 392-394.

27. Binner, J. G. P.; Hassine, N. A.; Cross, T. E., The Possible Role of the Preexponential Factor in Explaining the Increased Reaction-Rates Observed During the Microwave Synthesis of Titanium Carbide. *Journal of Materials Science* **1995**, 30, (21), 5389-5393.

Figures:

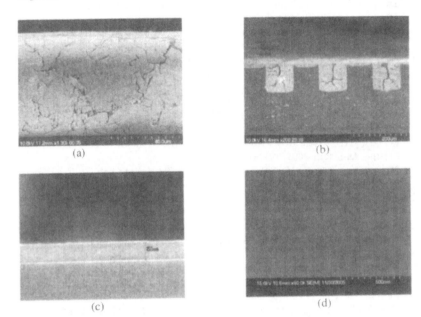

(a)

(b)

(c)

(d)

Figure 1

(a)

(b)

Figure 2

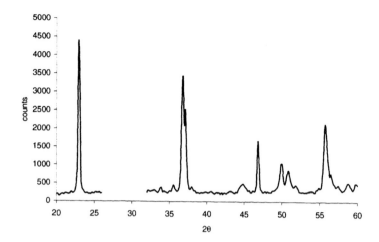

Figure 3

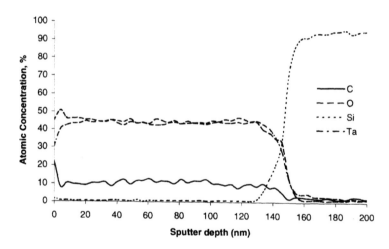

Figure 4

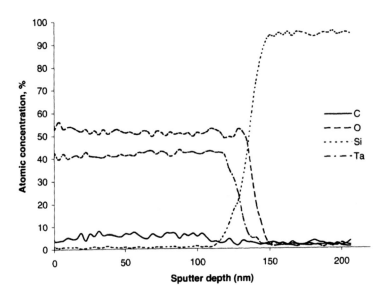

Figure 5

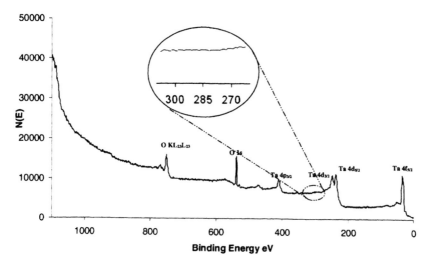

Figure 6

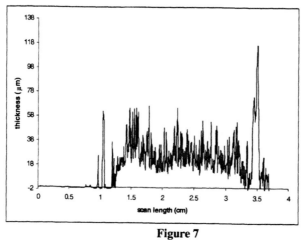

Figure 7

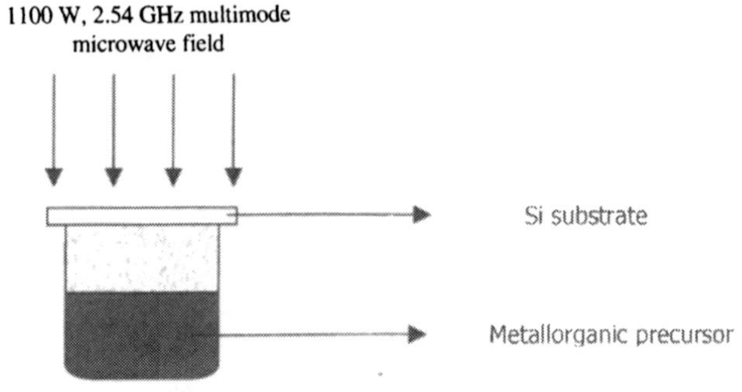

1100 W, 2.54 GHz multimode
microwave field

Si substrate

Metallorganic precursor

Figure 8

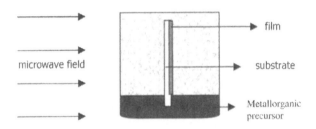

microwave field

film

substrate

Metallorganic
precursor

Figure 9

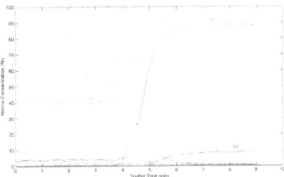

Figure 10

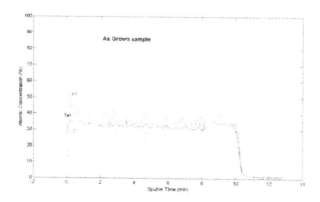

Figure 11

Figure captions:

Figure 1: SEM micrographs of
(a), cross sections of vertically oriented, unpolished Si substrate (7 min exposure), 62μm thick film. (b) cross section of vertically oriented, patterned Si substrate (7 min exposure)
(c), (d) horizontally oriented, polished Si substrate (10 sec exposure), 150 nm thick ((c) is cross section view, (d) is surface image).

Figure 2: SEM cross sections of calcined vs uncalcined Ta_2O_5 (horizontally oriented deposition, 1.5 min exposure)
(a) uncalcined Ta_2O_5 (800nm thick)
(b) calcined Ta_2O_5 (575 nm thick)

Figure 3: XRD spectrum of calcined Ta_2O_5

Figure 4: AES spectrum of uncalcined Ta_2O_5 (10sec exposure)

Figure 5: AES spectrum of calcined Ta_2O_5 (10sec exposure)

Figure 6: XPS spectrum of calcined film

Figure 7: Surface profile of horizontally oriented substrate

Figure 8: Deposition scheme: Horizontally oriented substrate

Figure 9: Deposition scheme: Vertically oriented substrate

Figure 10: AES spectrum of 130 nm uncalcined Ta_2O_5

Figure 11: AES spectrum of 800 nm uncalcined Ta_2O_5

Mater. Res. Soc. Symp. Proc. Vol. 929 © 2006 Materials Research Society   0929-II04-12

Improvement on Thermoelectric Characteristics of Layered Nanostructure by Ion Beam Bombardment

B. Zheng, S. Budak, C. Muntele, Z. Xiao, S. Celaschi, I. Mutele, B. Chhay, R. L. Zimmerman, L. R. Holland, and D. Ila

Center for Irradiation of Materials -Alabama A & M University, P.O.Box 1447, Normal, AL, 35762

ABSTRACT

We made an n-type nanoscale multilayer superlattice thermoelectric cooling device which consists of multiple periodic layers of Si $_{1-x}$ Ge $_x$ / Si, The thickness of each layer ranges between 10 and 50 nm. The super lattice was bombarded by 5 MeV Si ions with different fluence in order to form nanoscale cluster quantum dot structures. We measured the thin film cross plane thermal conductivity with 3ω method, measuring the cross plane Seebeck coefficient, and finally measuring the cross plane electric conductivity before and after ion bombardment. As predicted, the thermo-electric figure of merit of the films increases with the increase of Si ion fluence. In addition to the effect of quantum well confinement of the phonon transmission, the nano-scale crystal quantum dots produced by the incident Si ion beam further adversely affect the thermal conductivity. The defect and disorder in the lattice caused by bombardment and the grain boundary of these nano-scale cluster increase scattering of phonon and increase the chance of the inelastic interaction of phonon and annihilation of phonon, this limits phonon mean free path, phonons are chiefly absorbed and dissipated along the lattice, therefore reduces the cross plane thermal conductivity, The increases of the electron density of state in the miniband of nanoscale cluster quantum dot structure formed by bombardment also increases Seebeck coefficient, and the electric conductivity. Eventually, the thermo-electric figure of merit of the films increases.

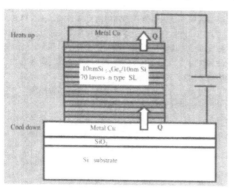

Fig. 1. Schematic of Si $_{1-x}$ Ge $_x$ / Si superlattice TE device

INTRODUCTION

Si / Si$_{1-x}$ Ge$_x$ multiplayer superlattice thin film thermo-electric (TE) cooling devices was made in CIM as shown in Fig. 1. This thin films is a periodic quantum well structure consisting of tens to hundreds of alternating layers with different band gap (Si: 1.12 eV, Si $_{1-x}$ Ge $_x$: 0.8~0.9 eV), The thickness of each layer ranges between 10 and 50 nm. The performance of superlattice thin film thermoelectric device is quantified by the dimensionless figure of merit $ZT = S^2 \sigma T / k$. Our aim is to obtain high ZT values by increasing the Seebeck coefficient S and the electrical conductivity σ, and reducing the thermal conductivity k by bombarding the superlattice structure with MeV Si ion. The bombardment tends to form nanoscale cluster of quantum dot structures. In addition to the quantum well confinement of the phonon transmission due to Bragg scattering and reflection at lattice interface [1, 2, 3], the defect and disorder in the lattice caused by bombardment and the grain boundary of these nanoscale cluster formed by bombardment increase the scattering of phonon and increase the chance of the inelastic interaction of phonon and the annihilation of phonon, inhabiting heat transport in the direction perpendicular to the lattice [4, 5, 6, 7]. Phonon is chiefly absorbed and dissipated along the lattice, so cross plane thermal conductivity will decrease.These quantum dot clusters also increase the Seebeck coefficient and electric conductivity due to the increase of the electronic density of states in nanoscale cluster 3D miniband in this quantum dot crystal with periodic potential barrier.

EXPERIMENTAL DETAILS

An electron beam deposition system with two guns was used to make the $Si_{1-x}Ge_x$ / Si, multilayer thin films. The chamber was pumped down by a cryogenic pump to a background pressure of 2×10^{-6} Torr. The multilayer films were sequentially deposited on Si substrate that was coated with a SiO_2 insulation layer and a metal (Cu) contact layer to form a periodic structure consisting of 70 alternating layers of $Si_{1-x}Ge_x$ / Si. In this quantum well superlattice structure (SL), The thickness of each layer ranges between 10 and 50 nm, A second Cu contact layer was later deposited over the SL after bombardment. The sample was cut by diamond cutter to get a clear edge of each layer. In this way, a complete cooling device was made. The thickness of the deposited layer was controlled by a crystal monitor. For each Si layer, the current of electron beam gun was set at 110 A, the rate of deposition was 5 Hz/sec. For each $Si_{1-x}Ge_x$ layer, the current of electron beam gun was set at 110 A for Si deposition and 40 A for Ge deposition respectively, the rate of deposition rate was 10 Hz/sec.
In order to determine the stoichiometry of $Si_{1-x}Ge_x$ layer grown in this condition, a single

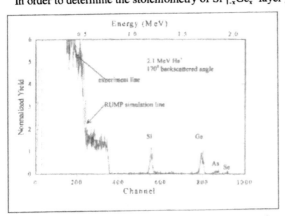

layer $Si_{1-x}Ge_x$ thin film was grown on glass polymeric carbon (GPC) for purpose of Rutherford backscattering spectroscopy (RBS) analysis. Figure 2 shows the RBS spectrum of a single $Si_{1-x}Ge_x$ layer grown on GPC and the RUMP (RBS Analysis and Simulation Package) simulation results. The RUMP simulation result indicates that the $Si_{1-x}Ge_x$ grown at this condition is characterized to be $Si_{0.84}Ge_{0.16}$.

Fig. 2. Rutherford Backscattering Spectroscopy (RBS) spectrum of a single layer $Si_{1-x}Ge_x$ thin film on a GPC and RUMP simulation results

The 5 MeV Si ion bombardments were performed using the AAMU Pelletron ion beam accelerator. The SRIM simulation software shows that the Si ions with the energy value of 5 MeV go through Si$_{1-x}$Ge$_x$ / Si multilayer superlattice film and terminate deeply in the substrate

THERMOELECTRIC CHARACTERISTIC MEASUREMENTS OF ION BOMBARDED MULTILAYER SUPERLATTICE

Cross plane thermal conductivity using 3ω method

In 3ω measurement of thermal conductivity [8] of thin film as shown in Fig 3. A Pt strip used as heater is deposited on the film [3], [4]. A power supply with the frequency of ω and constant voltage amplitude is used to heat the metal strip through the two terminals. The voltage signal is obtained through the other two terminals.

$$V = I_o \sin \omega t R_0 + \frac{\alpha I_o{}^3 R_0{}^2}{8kS}\sqrt{\frac{D}{\omega}}(\sin \omega t - \cos \omega t) + \frac{\alpha I_o{}^3 R_0{}^2}{8kS}\sqrt{\frac{D}{\omega}}(\cos 3\omega t - \sin 3\omega t) \qquad (1)$$

Where R_0 is sample electrical resistance, α is the temperature coefficient of resistance of Pt, S is the heat diffusion area between Pt strip and sample, D is the cross plane thin film diffusion coefficient, ρ is the density of the thin film, C is the specific heat of the thin film.

The 1 ω voltage signal component generated by the power supply is balanced off using a Kelvin bridge circuit. The smaller 2 ω signal caused by temperature variation will be culled off by a lock in amplifier. The remaining 3 ω signal caused by temperature variation will be selected by the lock-in amplifier [8, 9]. 3 ω signal voltage amplitude is:

$$V_s = \frac{\alpha I_o{}^3 R_0{}^2}{8kS}\sqrt{\frac{D}{\omega}} = \frac{\alpha I_o{}^3 R_0{}^2}{8S\sqrt{\omega C \rho}\sqrt{k}} \qquad (2)$$

So the thermal conductivity is:

$$k = \left[\frac{\alpha I_o{}^3 R_0{}^2}{8S\sqrt{\omega C \rho}V_s}\right]^2 \qquad (3)$$

Using data acquisition software Acquire and the whole spectrum frequency signal generated in the lock in amplifier itself to scan and display on computer the 3ω signal amplitude V_{31} induced by the first frequency f_1 signal and 3ω signal amplitude V_{32} induced by the second

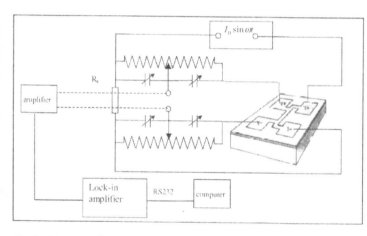

Fig. 3. Schematic of 3 ω measurement of thin film thermal conductivity

frequency f_2 signal, thermal conductivity can be given by formula [10, 11, 12]:

$$k = \frac{V^3 \ln \frac{f_2}{1}}{4\pi d R (V_{31} - V_{32})} \alpha \qquad (5)$$

Where l is the length of the strip.

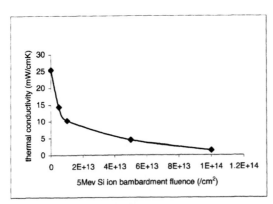

Fig.4. Thermal conductivity of 70 layers Si $_{1-x}$ Ge $_x$ / Si SL before and after 5 MeV Si ion bombardment

The thermal conductivity of 70 layer Si $_{1-x}$ Ge $_x$ / Si superlattice sample before and after bombardment by 5 MeV Si ion with fluency of 0.5×10^{13}/cm^2, 1 $\times 10^{13}$/cm^2, 5×10^{13}/cm^2, 10^{14}/cm^2 is shown in Fig. 4.

While at T = 300 K, thermal conductivity of Si is 1.48 W/mK. Fig.4 indicates that cross plane thermal conductivity decreases with increase of ion bombardment fluence.

Cross plane Seebeck coefficient

The experiment setup for measurement of Seebeck coefficient at various temperatures is shown in Fig. 5 (a). A reference wire (Constantan alloy) is used to eliminate the extraneous thermal offset voltage and to derive the Seebeck coefficient of the sample. At the cooling stage, Nitrogen cooling tubes are imbedded to cool down the sample and reference wire. In the sample holding stage, there is a heater (mean while serve as a thermometer) controlled by the temperature controller, which serves as modulation power supply to heat up hot end of the sample and reference wire to a certain temperature, to form temperature gradient between hot end and cold end of sample as well as reference wire. The heater resistor controlled by the temperature controller heats up the sample and the reference wire. The nitrogen cooling tubes and the heater resistor combine together to adjust the temperature of the sample and reference wire to a certain value when thermal equilibrium is reached. A silicon diode thermo sensor monitors the temperature at the hot end.

To measure the Seebeck coefficient of the sample at a certain temperature, all section of the sample and reference arm (Constantan reference material) is heated or cooled to this temperature, then using two different modulation power signal (0~900 mW) P_1, P_2 to heat up the hot end, to make the difference of temperature from the cold end to the hot end to be $\Delta T(P_1)$, $\Delta T(P_2)$. Seebeck coefficient of the sample S_s is calculated according to:

$$S_S = S_R \frac{V_S(P_2) - V_S(P_1)}{V_R(P_2) - V_R(P_1)} \qquad (6)$$

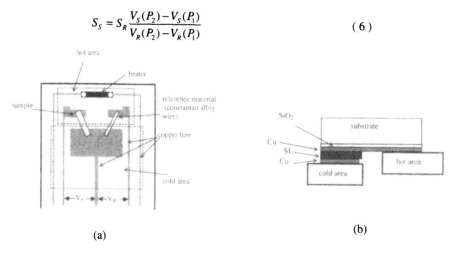

(a)

(b)

Fig. 5. (a) experiment setup for Seebeck coefficient measurement (b) sample for cross plane Seebeck coefficient measurement

S_R is the Seebeck coefficient of reference wire, $V_S(P_1)$, $V_S(P_2)$ $V_R(P_1)$ $V_R(P_2)$ is Seebeck effect voltage of sample and reference wire at different temperature gradient.

In order to get cross plane Seebeck coefficient of multiplayer supper lattice, the sample is prepared as shown in Fig. 5 (b), assuming that thermal conductivity of Cu is high enough to get heat equilibrium instantly.

The Seebeck coefficient of 70 layer Si/Si + Ge super lattice sample before and after bombardment by 5 MeV Si ion with fluence of $0.5 \times 10^{13}/cm^2$, $1 \times 10^{13}/cm^2$, $5 \times 10^{13}/cm^2$, $10^{14}/cm^2$ is shown in Fig. 6. While T=300K, the amplitude of the Seebeck coefficient of Si is 450 µV/K, the amplitude of Seebeck coefficient of Ge is 548 µV/K. According to Fig. 6, the amplitude of the Seebeck coefficient of the thin films increases with bombardment fluence. For n-type material, Seebeck coefficient

$$S = \frac{\Delta V}{\Delta T} = -\frac{k_B}{e} \ln \frac{N_C}{N_D}, \text{ where } N_c \text{ is}$$

effective density of state in conduction band,

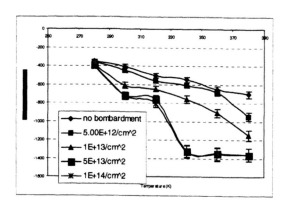

Fig. 6. Seebeck coefficient of 70 layers $Si_{1-x}Ge_x / Si$ SL (10 nm each layer) at different temperature before and after 5 MeV Si ion bombardment with different fluence

$$N_c = 2\left(\frac{2\pi m_c KT}{h^2}\right)^{\frac{3}{2}}$$, that is why Seebeck coefficient is proportional to temperature. The negative value of Seebeck coefficient indicates that our sample is n type.

Cross plane electric conductivity

Cross-plane electrical conductivity can be measured by contacting electrode of ammeter and voltmeter to the copper contact layers of the sample as shown in Fig. 7. Power supply is adjusted to 1 mV using potential divider to prevent penetrating the SL. Here we assume that Schottky junction barrier between Cu and semiconductor is negligible due to medium doped donor in the semiconductor material we used and resistance of copper layer is also negligible.

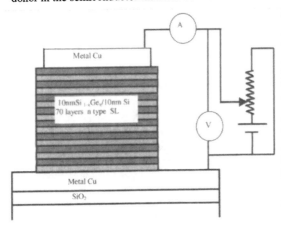

The electric conductivity of 70 layer $Si_{1-x} Ge_x$ / Si superlattice sample before and after bombardment by 5 MeV Si ion with fluency of $0.5 \times 10^{13}/cm^2$, $1 \times 10^{13}/cm^2$, $5 \times 10^{13}/cm^2$, $10^{14}/cm^2$ is shown in Fig. 8.

While at T =300K, intrinsic silicon electric conductivity is 4.3×10^{-6} $(\Omega cm)^{-1}$. Fig. 8 shows that electric conductivity increase with bombardment fluence.

The figure of merit (ZT= $S^2 \sigma T/k$) at different bombardment fluence is shown in Fig. 9, which indicates that the figure of merit of SL increases with the increase of bombardment fluence.

Fig. 7. Measurement of cross plane electric conductivity of thin film

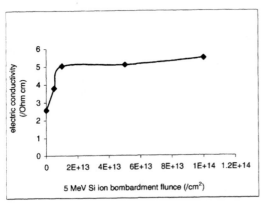

Fig.8. Cross plane electric conductivity of 70 layers $Si_{1-x} Ge_x$ / Si SL before and after 5 MeV Si ion bombardment with different fluence

DISCUSSION

Phonon transmission can be minimized as long as the superlattice period a is more than or equal to the half of phonon wavelength $2a \geq \lambda$, acoustic impedance mismatch between the materials of each layer give rise to selective Bragg scattering, and reflecting at periodic interface, if superlattice period satisfy the phonon Bragg scattering condition $2a\sin\theta = n\lambda$, here $\theta = \pi/2$, so phonon

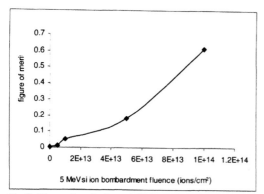

Fig. 9. Figure of merit of sample75 layers Si $_{1-x}$ Ge $_x$ / Si SL before and after 5 MeV Si ion bombardment with different fluence

is confined within the quantum well of superlattices (phonon blocking), however, the superlattice period can not be too large, otherwise the effect of superlattice will disappear, the acoustic wave will become the oscillation of atom instead of lattice. Because the distance between neighboring atoms a' is small, meaning $\lambda \geq 2a'$, so the phonon wavevector lies in the first Brillouin zone $k \leq \pi/a'$, phonon transmission can not be decreased.

Quantum well confinement effect is greatly enhanced by ion bombardment that formed nanocluster quantum dots, The defect and disorder in the lattice caused by bombardment and the grain boundary of these nanoscale cluster increase the scattering of phonon and increase the chance of the inelastic interaction of phonon and the annihilation of phonon, this limits phonon mean free path. This nanoscale cluster structure enhances phonon horizontal dissipation and absorption along the superlattice rather than perpendicular to superlattice, therefore decreases cross plane thermal conductivity. Although the disorder in the crystal resulted from bombardment in certain extent will decrease the electric conductivity, this tendency is overwhelmed by the increase of electronic density of state in the miniband of nanoscale cluster quantum dot structure formed by bombardment, electron transport perpendicular to the superlattice interface is enhanced eventually, These nano-scale clusters increase the Seebeck coefficient due to the increase of electronic density of states. So the figure of merit of the superlattice will increase. After bombardment, the superlattice crystal behaves more like an electronic crystal and phonon glass in the direction perpendicular to the superlattice interface. However, excessive bombardment will damage the structure of superlattice.

Besides, after bombardment, the alloy behave like diatomic crystal, because of the existence of nanocluster that is analogue to the heavy atom and together with original host atom that analogue to the lighter atom, so the phonon transmission is partially contributed by optical phonon in addition to acoustic phonon, the calculation of the specific heat of medium with optical phonon follow the Einstein model, which is smaller than the specific heat of medium with acoustic phonon that follow the Debye model, since k=1/3 (Cρv l_{mfp}), therefore thermal conductivity decreases. So Si $_{1-x}$ Ge $_x$ / Si semiconductor superlattice structures bombarded by Si ions have shown promise as high figure of merit thermoelectric materials.

REFERENCES

1. Xiaofeng Fan, Electronics Letters Jan 2001 Vol.37, No.2
2. Xiaofeng Fan, Gehong Zeng, Appl. Phys. Lett. Vol 78, No. 11, Mar. 2001
3. Rajeev Singh, Mat. Res. Soc. Symp. Proc. Vol. 793 © 2004 Materials Research Society
4. D.G. Cahill, M. Katiyar, Phys. Rev. B 50 (1994) 6007
5. Rama Venkatasubramanian, Phys. Rev. B 3091/ No. 4/Vol. 61, Jan. 2000

6. J. L. Liu, Physics Review B 67, 2003
7. Per Hyldgaard, Vol. 56, No. 17, Phys. Rev. B 1 Nov. 1997
8. David G. Cahill, Rev. Sci. Instrum. Vol. 61, No. 2, Feb. 1990
9. T. Borca-Tasciuc, Rev. Sci. Instrum. Vol. 72, No. 4 April 2001
10. D.G. Cahill, 134-142 / Vol 32, High Temperature –High Pressures 2001
11. S.M Lee, D.G. Cahill, Appl. Phys. Lett. 70 (22). 2 June 1997
12. Gehong Zeng, Appl. Phys. Lett. 71,1234-1236, 2004

Mater. Res. Soc. Symp. Proc. Vol. 929 © 2006 Materials Research Society   0929-II04-18

Sol-Gel Synthesis of Thick Ta$_2$O$_5$ Films for Photonic Band Gap Materials

Nicholas Ndiege[1], Tabitha Wilhoite[1], Vaidyanathan Subramanian[2], Mark Shannon[3], and Rich Masel[2]

[1]Chemistry, University of Illinois, Urbana, IL, 61801
[2]Chemical & Biomolecular Engineering, University of Illinois, Urbana, IL, 61801
[3]University of Illinois, Urbana, IL, 61801

Introduction

Insulation for high temperature microelectromechanical systems has become a key area in research due to the development of microreactor systems that can attain wall temperatures in the order of 1200°C.[1, 2] In order to add utility and portability to such systems it is required to thermally isolate parts of composite devices that incorporate such high temperature components. Refractory blocks (e.g. fibrous alumina) of centimeter order of thickness are the best mode of insulation currently available.

Photonic crystals have demonstrated a tunabilty such as to exhibit a photonic bandgap in the infrared wavelength.[3] This is promising since at temperatures above 600°C, IR radiation is the dominant mode of heat loss.[4, 5] Thus multilayer quarterwave photonic structures can be explored to not only effectively insulate but also to anisotropically expel heat generated by microreactors.

The greatest challenge posed in realizing such materials is that photonic crystals with a bandgap in the IR calls for structures made out of dense refractory material with thicknesses ranging from 1-5 µm.[3] Ta$_2$O$_5$ is the material of choice due to its high index of refraction, refractory nature and negligible absorbance in the infrared region.[4, 6, 7] Current deposition techniques typically attain thicknesses in the order of 1-1.5 µm.[7-14] Furthermore, these are achieved by use of high cost systems such as chemical vapour deposition, e-beam deposition etc. Any attempts at thicknesses beyond this range are hampered by very slow growth rates, film buckling, cracking and even catastrophic delamination.[9, 15, 16] Solgel synthesis techniques are appealing due to their low cost and flexibility with regards to chemical tuning of the properties of the final material sought. Binder molecules such as ethylcellulose, polyethyleneimine (PEI), polyvinyl pyrollidone (PVP) and polyvinyl butyral (PVB) have been shown to increase film thickness with little or no cracking when incorporated in the sol precursor.[17-22]

Previous studies have shown PVP to yield dense, 2.4µm thick Ta$_2$O$_5$ films on silicon with negligible cracking. Also observed are changes in film microstructure as the sol ages.[23] These films retained their stability even after prolonged heating at high temperature. Despite these discoveries, the mechanisms via which PVP and the corresponding solvent, facilitates the superior quality of oxide films still remains to be understood, hence in this paper, we seek to investigate how PVP interacts with its corresponding solvent and the tantalum complex in the preparation of thick Ta$_2$O$_5$ films. The sol precursors were analyzed with time from the point of synthesis till 12 days later. Chemical probing was performed via FTIR and NMR analysis. Thermogravimetric analysis (TGA) of the sol precursor was also performed. The resulting oxide films were characterized via scanning electron microscopy (SEM), X-Ray diffraction (XRD) and X-Ray Photoelectron Spectroscopy (XPS). The binder was found to have a viscosity increasing effect when incorporated into the sol precursor. This allowed for thicker films due to the increased surface tension during spin or dip coating. PVP also reduces the rate of solvent expulsion from the wet film via hydrogen bond formation. It also hydrogen bonds with the growing metal-oxide network, both of which limit crack formation during drying and calcination.[18, 24]

Experimental Section

Sol synthesis: 1-propanol (99+%) was added to tantalum (V) ethoxide (99.98%) in a 100ml round bottomed flask while stirring to prevent rapid hydrolysis of the latter when exposed to ambient moisture. PVP (avg mw 1, 300, 000) was weighed out in a beaker to which 2-methoxyethanol (99.3+%) was added while stirring (molecular weight of PVP monomer was used in the calculations). The mixture was further sonicated for 15 minutes to achieve complete binder dissolution. The PVP & 2-methoxyethanol solution was then added to the tantalum (V) ethoxide & 1-propanol solution while stirring. The resulting mixture was refluxed for 2 hours at 130°C. This was followed by vacuum evaporation until a 12% loss in mass was attained. The solution was then let to cool to room temperature, after which acetylacetone (99+%) was added, yielding a yellow colored solution. The solution was let to stir for 20 hrs in a capped flask. This resulting sol was used for deposition by spin coating at various age points. Further details are discussed elsewhere.[23] The films were then left to dry in air for 24 hrs then subjected to a 9hr bake at 900°C under an O_2 flow of 100 cm^3/min. Temperature ramping rate was 1.6°C/min for both up and down cycles.

Control samples were synthesized as above but with modifications as follows:
- Without PVP ($Ta(OC_2H_5)_5$, isopropanol, acetylacetone, 2-methoxyethanol)
- Without acetylacetone ($Ta(OC_2H_5)_5$, isopropanol, PVP, 2-methoxyethanol)
- Without 2-methoxyethanol (Acetylacetone as solvent) ($Ta(OC_2H_5)_5$, isopropanol, acetylacetone, PVP)

Results

FTIR analysis: Given the strong positively charged Ta metal center bonded to negatively charged ligands, the large dipole moment typically yields characteristically strong IR bands for metallorganic compounds. Infra red analysis also reveals the presence of hydrogen bonds.[25, 26] Sampling for FTIR was in the form of dip coated liquid films on a silicon substrate, analyzed on a Magna IR Spectrometer 750® from Nicolet Instrument Corp. FTIR in the reflectance mode at 45° incidence angle. Spectra of the solvent , PVP in 2-methoxyethanol solution and control sol without PVP were collected for reference purposes. Sampling of the Ta_2O_5 precursor sol and control sol without acetylacetone was conducted periodically starting at the end of sol synthesis, till the 12th day. Samples were Si <111> substrates (acetone-isopropanol rinsed and air dried) dip coated in the respective solutions at the indicated points in time using an in house dipcoater (withdrawal rate = 5 inches/min).

Analysis of sol with acetylacetone:

Spectrum is identical to that of 2-methoxyethanol except for C=O stretch at $1700 cm^{-1}$, $1297 cm^{-1}$ C-N stretch attributable to PVP molecule, and two bands at $650 cm^{-1}$ and $770 cm^{-1}$ attributable to Ta-O or Ta-O-C modes.[27, 29]

Analysis of control sol without acetylacetone:

Spectrum obtained is initially identical to 2-methoxyethanol except for $1297 cm^{-1}$ C-N and $1690 cm^{-1}$ C=O stretch modes due to PVP. With time, OH and methoxy CH_2 & CH_3 bands broaden out, leaving only the $1472 cm^{-1}$ CH_2 scissoring, $1690 cm^{-1}$ C=O and CH_2 bands at $\sim1100 cm^{-1}$ distinctly visible.[25, 26] Two bands at $650 cm^{-1}$ and $770 cm^{-1}$ attributable to Ta-O or Ta-O-C modes are present but broad and not as distinct as observed for the sol precursor.[27, 29] The sol visibly turns turbid with time. ^{13}C and ^{1}H NMR spectra taken over time do not show any change. The sol had solidified into a gel after the 12th day.

NMR analysis: ^{1}H and ^{13}C NMR analysis was performed using Varian UNITY 400® and Varian UNITY 500® systems. Wilmad 528-PP-7® sample tubes were used. This analysis was important in determining what ligands are attached to the metal center at the end of the synthesis, as well as the fate of the exchanged hydrocarbon groups.[30, 31] NMR sampling was conducted daily over a period of 18 days starting at the end of sol synthesis, and thereafter every 2 days till the 18th day from the end of synthesis. The NMR standard used was duterated chloroform 99.8 atom % D with 1 % (v/v) TMS. The different samples analyzed were:

- Sol synthesized above
- Control sol synthesized as above but without PVP
- All the above but with $CaSO_{4(s)}$ pellet acting as H_2O scavenger
- Control sol synthesized as above but without acetylacetone

There was no variation in the ^{1}H and ^{13}C spectra obtained from the ageing samples. Spectra obtained from the control samples were similar to the spectra of the actual precursor.

SEM analysis: Scanning Electron Microscopy (SEM) was performed on samples of green (uncalcined) films to examine the microstructure resulting from the sols of different ages. Samples of calcined films from control sols (acetylacetone free and 2-methoxyethanol free) were also imaged. Analysis was performed on JEOL 6060LV® and HITACHI S4700® SEM machines.

XPS analysis: X-ray photoelectron spectroscopy was used to investigate the residual carbon content of the Ta_2O_5 films after calcinations. Analysis was done on a PHI 5400® XPS system (Mg Kα x-ray source). The oxide upper surface and interior of the film (surface was sputtered with Argon for 15 minutes) were both analyzed. XPS was also used to reveal the Ta:O:C atomic ratio (table 2).

X-ray diffraction analysis: Identification of the phases of the resulting oxide film was performed via XRD on a Rigaku DMAX® XRD system with Cu (Kα) radiation.

TGA analysis: Thermogravimetric analysis was performed to elucidate the change in physical properties that occurs through the calcination cycle to yield the final oxide film. Analysis was performed on a TGA7/DSC7 Perkin Elmer® system. Samples were heated from ambient temperature to 900°C at a rate of 1.6°C/min under a steady O_2 flow of 20cm³/min and finally held at 900°C for 10 hours.

Discussion

The ^1H spectrum is of a characteristic 2-methoxyethoxide type of coordination[30, 31]. The triplet at 3.7 ppm on the ^1H spectrum corresponds to a CH_2 hydrogen of a carbon bound to another CH_2 whose hydrogen signal appears at about 3.48ppm (triplet) and is bound to an O atom.[30] The singlet at 3.38ppm corresponds to CH_3 hydrogen bound to an oxygen atom. The ^{13}C spectrum confirms the above coordination with the methyl carbon at 58.8ppm, the oxygen bonded CH_2 carbon at 73.6ppm and the adjacent CH_2 carbon at 61.7 ppm.[30] NMR spectra of samples with $CaSO_4$ pellet yielded identical spectra to the ones above, thus showing no artifacts due to water present in the sample.

The NMR spectra suggest that the ethoxide ligands around the Ta metal center are exchanged for the $-OCH_2CH_2OCH_3$ ligand. There is no trace of any ethoxide/ethanol, hence we can conclude that the exchanged ethoxide ligands are removed as ethanol during the vacuum evaporation stage of synthesis. The suggested product is $Ta(OCH_2CH_2OCH_3)_x$ via the reaction:[4, 9, 32-34]

$$Ta(OCH_2CH_3)_{5(l)} + xHOCH_2CH_2OCH_{3(l)} \rightarrow Ta(OCH_2CH_2OCH_3)_{x(l)} + 5HOCH_2CH_{3(l)}$$

^1H NMR spectrum of 2-methoxyethanol yields a quartet centered at 3.7 ppm. The occurrence of a triplet for this sample suggests the existence of deprotonated OH group possibly due to extensive hydrogen bonding. Presence of the Ta-O and Ta-O-C bonds is verified by the FTIR bands < 1000 cm⁻¹.

Analysis of sols at different points in time via both NMR and FTIR does not show any change in chemical composition of the sol. This suggests that the difference observed in film thickness and microstructure from sols of different age is either a physical effect, namely the binder interaction with the Ta complex (or the Ta metal center) and solvent, or a modification of the binder molecular chain arrangement relative to the growing M-O-M network. PVP is capable of hydrogen bonding via the C=O functional group, though findings in this work suggest that there could be more than H-bonding taking place. This is supported by the microstructure observed in the green films resulting from fresh sols. The polycondensed oxide particles are about 822 nm in diameter. As time goes by, there is further polycondensation of the oxide network giving larger particle sizes in the films derived from sols subjected to greater ageing (933 nm). SEM micrographs of green films from aged sols show particle sizes of about 1.11μm diameter. The calcined Ta_2O_5 films also echo this variation in microstructure.[23]

The sol precursor exhibits FTIR bands at 1603 and 1540 cm⁻¹ that are absent from the control and other comparison spectra. These bands fall within the region where Ta coordinated acetylacetanate modes occur and also have the expected 63cm⁻¹ separation.[27] This suggests that there are mixed ligands around the Ta center, though the acetylacetanate ligands are undetectable by NMR experiments, implying that they are significantly H-bonded to the PVP or to the growing M-O-M network. Therefore, we can conclude that the actual ligand exchange reaction that occurs during the sol synthesis is as follows:

$$Ta(OCH_2CH_3)_{5(l)} + xHOCH_2CH_2OCH_{3(l)} \rightarrow Ta(OCH_2CH_2OCH_3)_{x\,(l)} + 5HOCH_2CH_{3(l)}$$

$$Ta(OCH_2CH_2OCH_3)_{x\,(l)} + yC_5H_8O_{2(l)} \rightarrow Ta(OCH_2CH_2OCH_3)_{x-y}(C_5H_8O_2)_{y\,(l)} + 2yHOCH_2CH_{3(l)}$$

General polycondensation sequence of reaction is as follows:[4, 33, 34]

Hydrolysis: $2\ M(OR)_4 + 2\ H_2O \rightarrow 2\ (RO)_3M\text{-}OH + 2\ ROH$

Subsequent condensation reactions involving the hydroxyl groups yield networks composed of inorganic oxide (M-O-M) linkages.

Olation: $2\ (RO)_3M\text{-}OH \rightarrow (RO)_3M\text{-}O\text{-}M(OR)_3 + H_2O$

Oxolation: $(RO)_3M\text{-}OH + M(OR)_4 \rightarrow (RO)_3M\text{-}O\text{-}M\ (RO)_3 + ROH$

The extensive H-bonded network made up of PVP, solvent and hydrolysis product allows for thicker films than those derived from binder-free sols due to the increased mechanical tolerance. The mixed ligands present results in a directionally different rate of hydrolysis around the metal center with the acetylacetanate groups being slower to remove due to their charge stabilizing effect. The resulting difference in growth rates is likely to force the M-O-M network formation to take on the globular microstructure. This type of microstructure is absent in films deposited from the acetylacetone free sol. With time, the acetylacetanate ligands undergo polycondensation as well; albeit at a much slower rate, hence, the denser, thinner films with larger spheroidal particles. The calcined films derived from acetylacetone-free sol have much smaller particles of fractal nature. On the other hand, sols comprised only of acetylacetone as solvent with PVP and tantalum (v) ethoxide do not show the spherical microstructure either. Acetylacetone has been observed in other studies to not only slow down the rate of hydrolysis and polycondensation but also to significantly influence the morphology and crystallinity of the final product of the sol gel synthesis.[35, 36]

TGA results show an initial drop in mass up to 50°C attributed to solvent evaporation. Between 60°C and 500°C there is a change in the rate of decrease in % weight, with the PVP assisted sample showing a slower rate of change than the sample without PVP. There is a significant increase in the rate of decrease of % weight at 450 °C, which is attributable to a change in structure from amorphous to orthorhombic Ta_2O_5. Beyond 500°C the rate of change in % weight is similar for both samples. This supports the hypothesis that the presence of PVP serves to relieve stresses within the film upon thermal treatment up to about 500°C when all the binder is pyrolised.[22, 28] The slower rate of decrease in % weight signifies slower change in the film microstructure, a plasticity that allows differential expansion between the film particles and substrate with minimal or no cracking of the film.

XPS data reveals 13.8% carbon content that is reduced to 0.9% upon sputtering with argon, confirming ambient contamination as the carbon source. This shows that the removal of PVP is complete and there is negligible residual carbon. The Ta:O ratio confirms atomic coordination as Ta_2O_5.

XRD pattern identifies the calcined film obtained as orthorhombic Ta_2O_5. The spectrum matches ICSD card # 97-004-7493.

Conclusion

PVP has been shown to be an effective binder material in the sol gel synthesis of thick, crack free Ta_2O_5 films for high temperature applications. This study was undertaken to investigate the effect of PVP interaction with the solvent and metal complex in the sol precursor via NMR, FTIR, TGA, XPS, SEM, and XRD analytical tools. It has been reported that the type of Ta_2O_5 films obtained from sols of different ages vary in thickness and microstructure, suggesting that there was a progressive change occurring in the composition of the sol precursor with time. This is due to different particle growth rates around metallic centers during Ta—O—Ta network formation resulting in spherical particles. This type of microstructure hydrogen bonds with the PVP matrix in a manner that effectively relieves stresses that occur in the film during solvent removal and calcination processes. The resulting films are found to be stable and crack free with maximum thickness of films deposited from freshly prepared sols. With time, there is continued polycondensation of the Ta—O—Ta network yielding denser, thinner films with particles of larger diameter.

Acknowledgments: SEM, XRD and XPS analysis was carried out in the Center for Microanalysis of Materials, University of Illinois, which is partially supported by the U.S. Department of Energy under grant DEFG02-91-ER45439

References:

1. Ganley, J. C.; Seebauer, E. G.; Masel, R. I., Porous anodic alumina microreactors for production of hydrogen from ammonia. *Aiche Journal* **2004**, 50, (4), 829-834.
2. Ganley, J. C.; Thomas, F. S.; Seebauer, E. G.; Masel, R. I., A priori catalytic activity correlations: the difficult case of hydrogen production from ammonia. *Catalysis Letters* **2004**, 96, (3-4), 117-122.
3. Fleming, J. G.; Lin, S. Y.; El-Kady, I.; Biswas, R.; Ho, K. M., All-metallic three-dimensional photonic crystals with a large infrared bandgap. *Nature* **2002**, 417, (6884), 52-55.
4. Hass, D. D. P., D.; Glass, D. E.; Wiedemann, K. E *Reflective Coating on Fibrous Insulation for Reduced Heat Transfer*; August, 1997; p^pp.
5. Viskanta, R., Overview of convection and radiation in high temperature gas flows. *International Journal of Engineering Science* **1998**, 36, (12-14), 1677-1699.
6. Mir, J. M.; Agostinelli, J. A., Optical Thin-Films for Wave-Guide Applications. *Journal of Vacuum Science & Technology a-Vacuum Surfaces and Films* **1994**, 12, (4), 1439-1445.
7. Ozer, N.; Lampert, C. M., Structural and optical properties of sol-gel deposited proton conducting Ta2O5 films. *Journal of Sol-Gel Science and Technology* **1997**, 8, (1-3), 703-709.
8. Ezhilvalavan, S.; Tseng, T. Y., Preparation and properties of tantalum pentoxide (Ta2O5) thin films for ultra large scale integrated circuits (ULSIs) application - A review. *Journal of Materials Science-Materials in Electronics* **1999**, 10, (1), 9-31.

9. Chaneliere, C.; Autran, J. L.; Devine, R. A. B.; Balland, B., Tantalum pentoxide (Ta2O5) thin films for advanced dielectric applications. *Materials Science & Engineering R-Reports* **1998**, 22, (6), 269-322.

10. Beinhorn, F.; Ihlemann, J.; Simon, P.; Marowsky, G.; Maisenholder, B.; Edlinger, J.; Neuschafer, D.; Anselmetti, D., Sub-micron grating formation in Ta2O5-waveguides by femtosecond UV-laser ablation. *Applied Surface Science* **1999**, 139, 107-110.

11. Kukli, K.; Aarik, J.; Aidla, A.; Kohan, O.; Uustare, T.; Sammelselg, V., Properties of Tantalum Oxide Thin-Films Grown by Atomic Layer Deposition. *Thin Solid Films* **1995**, 260, (2), 135-142.

12. Corbella, C. V., M.; Pinyol, A.; Porqueras, C.; Person, C.; Bertran, E., Influence of the porosity of RF sputtered Ta2 O5 thin films on their optical properties for electrochromic applications. *Solid State Ionics* **2003**, 165, 15-22.

13. Toki, K.; Kusakabe, K.; Odani, T.; Kobuna, S.; Shimizu, Y., Deposition of SiO2 and Ta2O5 films by electron-beam-excited plasma ion plating. *Thin Solid Films* **1996**, 282, (1-2), 401-403.

14. Werder, D. J.; Kola, R. R., Microstructure of Ta2O5 films grown by the anodization of TaNx. *Thin Solid Films* **1998**, 323, (1-2), 6-9.

15. Joshi, P. C.; Cole, M. W., Influence of postdeposition annealing on the enhanced structural and electrical properties of amorphous and crystalline Ta2O5 thin films for dynamic random access memory applications. *Journal of Applied Physics* **1999**, 86, (2), 871-880.

16. Liu, L.; Wang, Y.; Gong, H., Annealing effects of tantalum films on Si and SiO2/Si substrates in various vacuums. *Journal of Applied Physics* **2001**, 90, (1), 416-420.

17. Kozuka, H.; Takenaka, S.; Tokita, H.; Okubayashi, M., PVP-assisted sol-gel deposition of single layer ferroelectric thin films over submicron or micron in thickness. *Journal of the European Ceramic Society* **2004**, 24, (6), 1585-1588.

18. Kozuka, H.; Kajimura, M.; Hirano, T.; Katayama, K., Crack-free, thick ceramic coating films via non-repetitive dip-coating using polyvinylpyrrolidone as stress-relaxing agent. *Journal of Sol-Gel Science and Technology* **2000**, 19, (1-3), 205-209.

19. Chen, W.; Zhang, J. Y.; Fang, Q.; Li, S.; Wu, J. X.; Li, F. Q.; Jiang, K., Sol-gel preparation of thick titania coatings aided by organic binder materials. *Sensors and Actuators B-Chemical* **2004**, 100, (1-2), 195-199.

20. Chen, Y. Y.; Wei, W. C. J., Formation of mullite thin film via a sol-gel process with polyvinylpyrrolidone additive. *Journal of the European Ceramic Society* **2001**, 21, (14), 2535-2540.

21. Jia, Q. X.; Mccleskey, T. M.; Burrell, A. K.; Lin, Y.; Collis, G. E.; Wang, H.; Li, A. D. Q.; Foltyn, S. R., Polymer-assisted deposition of metal-oxide films. *Nature Materials* **2004**, 3, (8), 529-532.

22. Kishimoto, T.; Kozuka, H., Sol-gel preparation of TiO2 ceramic coating films from aqueous solutions of titanium sulfate (IV) containing polyvinylpyrrolidone. *Journal of Materials Research* **2003**, 18, (2), 466-474.

23. Subramanian, V. N., N.; Seebauer, E. G.; Shannon, M. A.; Masel, R. I., Synthesis of PVP assisted Ta2O5 Films and its Characterization. *Thin Solid Films* **2005**, in press.

24. Fidalgo, A.; Ilharco, L. M., Thickness, morphology and structure of sol-gel hybrid films: I - The role of the precursor solution's ageing. *Journal of Sol-Gel Science and Technology* **2003**, 26, (1-3), 363-367.

25. Smith, B., *Infrared Spectral Interpretation: A systematic Approach*. ed.; CRC Press: Boca Raton, Florida (USA), 1999; 'Vol.' p.

26. Coates, J., Interpretation of Infrared Spectra, A Practical Approach. In *Encyclopedia of Analytical Chemistry*, ed.; Meyers, R. A., 'Ed.'^'Eds.' John Wiley & Sons, Ltd: Chichester, 2000; 'Vol.' p^pp 10815-10837.

27. Phule, P. P., Sol-Gel Synthesis of Ferroelectric Lithium Tantalate Ceramics - Ftir Investigation of the Molecular Modification of Tantalum Ethoxide. *Journal of Materials Research* **1993**, 8, (2), 334-338.

28. Kozuka, H.; Higuchi, A., Single-layer submicron-thick BaTiO3 coatings from poly(vinylpyrrolidone)-containing sols: Gel-to-ceramic film conversion, densification, and dielectric properties. *Journal of Materials Research* **2001**, 16, (11), 3116-3123.

29. Kelly, P. V.; Mooney, M. B.; Beechinor, J. T.; O'Sullivan, B. J.; Hurley, P. K.; Crean, G. M.; Zhang, J. Y.; Boyd, I. W.; Paillous, M.; Jimenez, C.; Senateur, J. P., Ultraviolet assisted injection liquid source chemical vapour deposition (UVILS-CVD) of tantalum pentoxide. *Advanced Materials for Optics and Electronics* **2000**, 10, (3-5), 115-122.

30. Kim, Y.; Chae, H. K.; Lee, K. S.; Lee, W. I., Preparation of SiBi2Ta2O9 thin films with a single alkoxide sol-gel precursor. *Journal of Materials Chemistry* **1998**, 8, (11), 2317-2319.

31. Werndrup, P.; Verdenelli, M.; Chassagneux, F.; Parola, S.; Kessler, V. G., Powders and dense thin films of late transition metal oxide nanocomposites from structurally characterized single-source precursors. *Journal of Materials Chemistry* **2004**, 14, (3), 344-350.

32. Clem, P. G.; Jeon, N. L.; Nuzzo, R. G.; Payne, D. A., Monolayer-mediated deposition of tantalum(V) oxide thin film structures from solution precursors. *Journal of the American Ceramic Society* **1997**, 80, (11), 2821-2827.

33. Brinker, C. J.; Scherer, G. W., Sol-gel science-the physics and chemistry of sol-gel processing. *Academic press, inc.1250 sixth avenue,San Diego,CA,92101* **1990**.

34. Hubert-Pfalzgraf, L. G., Toward molecular design of homo and heterometallic precursors of lanthanide oxide-based materials. *New J. Chem* **1995**, 19, 727- 750.

35. Nakagawa, K.; Wang, F. M.; Murata, Y.; Adachi, M., Effect of acetylacetone on morphology and crystalline structure of nanostructured TiO2 in titanium alkoxide aqueous solution system. *Chemistry Letters* **2005**, 34, (5), 736-737.

36. Maslosh, V. Z.; Kotova, V. V.; Maslosh, O. V., Effect of acetylacetone on the residual content of formaldehyde in urea-formaldehyde resin. *Russian Journal of Applied Chemistry* **2002**, 75, (8), 1369-1370.

Diagnostics and Novel
Characterization Techniques

Mater. Res. Soc. Symp. Proc. Vol. 929 © 2006 Materials Research Society   0929-II03-01

A Novel Mechanical Method to Measure Shear Strength in Specimens Under Pressure

Juan Pablo Escobedo[1], David Field[1], David Lassila[2], and Mary Leblanc[2]
[1]Mechanical and Materials Engineering, Washington State University, Spokane and College, Pullman, WA, 99164
[2]Lawrence Livermore National Laboratory, Livermore, Ca, 94550

ABSTRACT

A new experimental apparatus has been developed for performing shear tests on specimens held under moderately high hydrostatic pressures (on the order of 4 GPa). This testing procedure experimentally determines the pressure-dependent shear strength of thin foil specimens. The experiments provide calibration data for models of materials subjected to extreme pressures such as the Steinberg-Guinan hardening model and can assist in model validation for discrete dislocation dynamics simulations, among others. This paper reports the development of the experimental procedures and the results of initial experiments on thin foils of polycrystalline Ta performed under hydrostatic pressures ranging from 1 to 4 GPa. Both yielding and hardening behavior of Ta are observed to be sensitive to the imposed pressure.

INTRODUCTION

Plastic deformation in metallic systems occurs primarily via dislocation generation and movement due to shear stresses. In the case of hydrostatic loading no shear stresses are present to provide a driving force for dislocation motion, therefore structural properties are not changed due to hydrostatic pressure alone. Nevertheless it is found in the literature that properties such as hardening and ductility of metals are sensitive to superimposed pressures. Even at relatively low pressures of 0.7-3.0 GPa a remarkable increase in ductility of some materials has been reported [1]. Bridgman's results [2,3,4,5] suggest that pressure hardening occurs in Mo, Ta and Ni.

Another property of interest is the shock-induced phase transformation. Numerous cases of phase changes have been reported when materials were subjected to pressures exceeding 30 GPa [6,7,8]. Finally from the physics perspective, the relationship between material strength, elastic constants, and microstructure is of great interest, and can lead to new insights of mechanisms of plastic flow under conditions of imposed pressure.

Most high-pressure research has been carried out under static conditions. For pressures in the range of 0-3 GPa testing has been conducted using a variety of media including solid, liquids and gases [1]. To achieve higher pressures, experiments have been conducted using the diamond anvil cell, [6], where the specimen is loaded to high pressures between the diamond anvils. Although this device allows ultrahigh pressures to be reached readily, it has the deficiency that the hydrostatic, frictional and deviatoric stresses increase in an uncontrolled manner as the load increases, making it difficult or impossible to determine strength and work hardening from experimental results. Another disadvantage of this method is that typically the volume of material tested in these kinds of systems is small and the properties that are observed may be functions of sample size. Postmortem analysis in these types of experiments is often difficult because of the small specimens necessarily used.

THEORY

In order to overcome these inconveniences a new test procedure, similar to that established by Bridgman [2-5], has been developed to investigate materials response in a pure shear deformation under high pressure conditions. This system is depicted in Fig 1. This test emphasizes two major features: strict control in the loading path, in order to separate the effects of hydrostatic and deviatoric stresses, and the ability to perform high pressure experiments using a larger specimen size that can be analyzed using standard characterization tools subsequent to testing, such as hardness measurements and TEM analysis.

In this modified Bridgman cell, three independent supported tungsten carbide anvils were used (figure 1 b). The anvils in the top platen are formed with a hemispherical section that fits snugly into the mating surface. Using a thin foil of indium between these hemispherical surfaces provides for self-leveling of the anvils upon initial loading. The same pressure is attained between all anvils by positioning each anvil atop a hydraulically-controlled piston, all of which are connected to a common oil reservoir in order to establish equivalent load on all anvils. Extensometers were attached to the anvils to measure the local strain on the sample being deformed and to avoid issues of machine compliance.

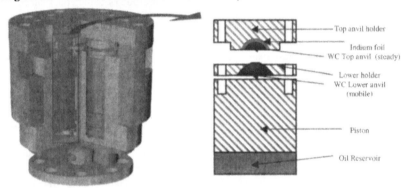

Figure 1. Schematic of the modified Bridgman cell used to perform the experiments.

The deformation of the specimens in the tri-anvil apparatus is achieved with the same general philosophy as with the original Bridgman design as shown in figure 1: axial load followed by rotational displacement.

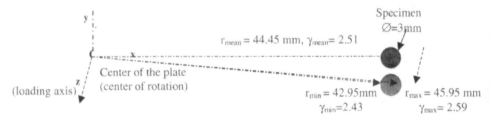

Figure 2. Top view of the arrangements of the anvils in the lower plate.

In this new experiment, the anvil centers are positioned on a circle centered on the loading axis of the testing system (figure 2). This arrangement produces a near negligible shear strain gradient, in fact this shear loading can be assumed unidirectional because of the small specimen size, (radius/thickness ratio ranges from 60 – 600) and low angle rotations ($\theta \approx 0.5^\circ = 8.7$ mrad) in comparison to the circumference of the circle on which the specimens lie $R \approx 44.45$ mm, a rough calculation leads to a value for the strain gradient of 6 % from end to end along the specimen thereby creating a more uniform shear stress in the specimen as compared with the original Bridgman single anvil design where the strain is radial function from the center of the specimen.

DISCUSSION

For this experiment, it is crucial to determine whether the desired state of hydrostatic pressure was be attained by application of axial load on the thin foil specimens between the two anvils. The aim is to have the major portion of the specimen under a uniform hydrostatic pressure. By using thinner foils, a more uniform distributed pressure can be achieved, but if the thickness is too small, the possibility exists that contact could occur between the anvils during specimen testing thereby confusing the measurement of load on the specimen. Assuming proper platen alignment, this contact could occur by either of two processes: elastic deformation of the anvils due to bearing stress, or specimen conformity to the roughened anvil surfaces.

To experimentally find the optimum specimen thickness, a set of Ta foils purchased from a commercial vendor and vacuum annealed at 1000C with thicknesses of 50, 25, 12.5 and 5 µm were used for the initial experiments described herein, the samples were disks of 3 mm in diameter. By using specimens of this size, a complete post-mortem characterization was possible in order to validate the adequacy of the testing device. The stress-strain response is shown next in figure 3.

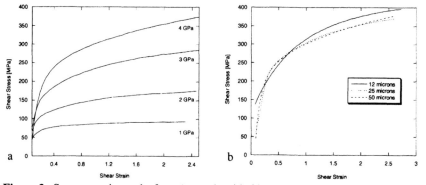

Figure 3. Stress – strain results for: a) sample with thickness of 50 microns, b) response of specimens with different thicknesses.

The samples with thicknesses of 25 and 12.5 microns show a similar response as the 50 micron samples, as it is depicted in figure 3b for a pressure of 4 GPa, the stress-strain response converged being independent on the thickness of the sample. The results of the 5 microns thick sample have been determined to be erroneous, it was established that these samples caused platen contact and binding due to elastic strain in anvils and test samples added to the inherent roughness of the anvils.

Hardening

A second issue imperative to the establishment of this new experimental technique is the determination that the stress state in the material upon initial loading is primarily hydrostatic. As it has been mentioned plastic deformation occurs primarily via dislocation generation and movement due to shear stresses. In the case of hydrostatic loading no shear stresses are present to provide a driving force for dislocation motion, therefore structural properties are not changed due to hydrostatic pressure alone. In addition, dislocation activity causes strain hardening in the material, so if significant dislocation activity occurs, the hardness of the metal after loading is expected to increase. To investigate this phenomenon, hardness measurements were made across the specimens on the different thicknesses subjected to pressures of 4 GPa, the maximum attained during the tests. The results of the 12.5 microns samples are shown in figure 4.

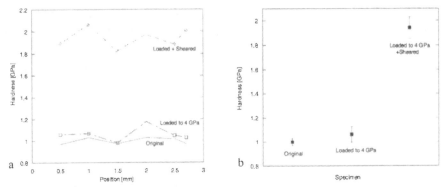

Figure 4. Hardness of specimens under 4.2GPa.

Hardness measurements were made every 0.5 mm from side-to-side across the specimen diameter. It is evident from figure 4 that the samples subjected to loading show no hardening effects due to the superimposed pressure, the original value was estimated to be 1.02±0.03 GPa, and after loading equal to 1.08± 0.04GPa, the two values being the same within the experimental error when measuring hardness. On the contrary, the samples sheared after loading show values of 1.97± 0.09 GPa, an evident increase in hardness close to 82%. These values of hardness corroborate that the stress-strain response is a measure of the strength of the specimens held under pressure and not an effect of *a priori* deformation caused by axial loading.

TEM analysis

Bright field TEM was done in order to reveal microstructural changes. This analysis was done on both the deformed and undeformed specimens. Figure 5a shows the image of the original Ta foil that was used for the tests. No appreciable dislocation density prior to the deformation process is observed. For the specimen loaded/unloaded to a pressure of 4.2 GPa without shearing (figure 5b), a slight increase in dislocation content is observed, but the grains are still relatively free from dislocation debris. This explains why the measurements of the hardness in the specimens loaded and unloaded without shearing were close to those obtained for the undeformed specimens.

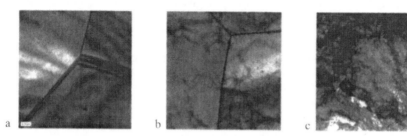

Figure 5. TEM analysis of the samples at 4.2 GPa. a) Original specimen, b) specimen loaded and unloaded with no shearing, and c) specimen loaded and sheared.

These results show that while some dislocation motion occurs near the specimen centers during loading of the specimens, the crystallites remain largely undeformed. The structure therefore must undergo primarily elastic strain during loading to high pressures. As for the sheared specimens it is evident that dislocation activity took place (figure 5c) and the mechanical properties were changed as evidenced by the increase in hardness.

CONCLUSIONS

The details of a new procedure to study the mechanical properties of materials deformed by shearing strains while maintained under high pressure have been described. Based on the results exposed here, this procedure has proven to be a good method to study the shear stress – shear strain behavior under hydrostatic pressure. It was observed that the microstructure and properties do not change significantly due solely to the effect of the pressure applied. This was corroborated via Vickers hardness characterization, with the hardness being almost equal to that of the un-deformed material. In addition, validation was done through structural analysis via TEM imaging. Neither of these techniques revealed significant deformation of the microstructure nor dislocation multiplication during the loading process.

The outcome of the tests showed a more influent role of the hydrostatic pressure on properties such as yield strength and hardening than that predicted by the Steinberg-Guinan model [9,10]. The model predicts an increment in yield strength of close to 3% up to a pressure of 4.2 GPa. Experimentally, an increase close to 180 % was observed for this material at the same pressure. These results are in qualitative agreement with those obtained by Bridgman and by Weir [11]. A more extensive analysis will be given in a later paper.

ACKNOWLEDGMENTS

Financial support from Lawrence Livermore National Laboratory is greatly appreciated.

REFERENCES

1. J.L. Lewandowski and P. Lowhaphandu, "Effect of hydrostatic pressure on mechanical behaviour and deformation processing of materials on materials", Int. *Mat. Reviews*, v **43**, n 4:145-187, 1998.
2. PW Bridgman, "Shearing phenomena at high pressures", Proc. Am. Acad. Arts Sci. **71**: 387-460, 1937.
3. Bridgman, P.W. 1935 "Effects of shearing stresses combined with high hydrostatic pressure", Phys. Rev. **48**: 825-847 (1935)
4. PW Bridgman, "Effects of hydrostatic pressure on the plastic properties of the metals", Rev. Mod. Phys **17**: 3-14, 1945.
5. PW Bridgman, "Flow phenomena in heavily stressed metals", J. Appl. Phys. **8**: 328-336, 1937.
6. Hsiung and D.H. Lassila, "Shock-induced deformation twinning and omega transformation in tantalum and tantalum-tungsten alloys," Acta Mater. **48**:4851-4865, 2000.
7. P.Söderlind and J.A. Moriarty, "First-principles theory of Ta up to 10 Mbar pressure: Structural and mechanical properties," Phys. Rev. B **57**:10340-10350, 1998.
8. Lassila, "Strength of materials under high pressure", Report LLNL.
9. DJ Steinberg, SG Cochran, MW Guinan "Constitutive model for metals applicable at high strain rate", J. Appl. Phys. **51**: 1498- 1505, 1980.
10. D. Steinberg, D Breithaupt, C Honodel "Work-hardening and effective viscosity of solid beryllium", Physica **139 & 140B**: 762-765 (1986).
11. S.R. Weir, J. Akella, C. Ruddle, T Goodwin and L. Siung. "Static strength of Ta and U under ultrahigh pressures." Phys. Rev. B **57**:11258-11265, 1998

Mater. Res. Soc. Symp. Proc. Vol. 929 © 2006 Materials Research Society   0929-II03-03

Surface Characterization of Silicon Carbide Following Shallow Implantation of Platinum Ions for High Temperature Hydrogen Sensing Applications

Claudiu Muntele, Satilmis Budak, Iulia Muntele, and Daryush Ila

Physics, Alabama A&M University, 4900 Meridian Street, PO Box 1447, Normal, AL, 35762

Abstract

Silicon carbide is a promising wide-bandgap semiconductor intended for use in fabrication of high temperature, high power, and fast switching microelectronics components running without cooling. For hydrogen sensing applications, silicon carbide is generally used in conjunction with either palladium or platinum, both of them being good catalysts for hydrogen. Here we are reporting on the temperature-dependent depth profile modifications of tungsten electrical contacts deposited on silicon carbide substrates.

Introduction

Palladium and platinum are known as good catalysts for hydrogen, which makes them good active materials in designing hydrogen sensors. They are absorbing hydrogen from the surrounding environment and then release it back in a continuous process, since they do not form stable hydrides. A typical hydrogen sensor design would consist of an electrically insulating substrate on which a thin Pd or Pt layer is deposited along with electrical contacts for monitoring changes in its resistivity due to hydrogen absorption. This concept is rather crude and not very sensitive, therefore the insulating substrate is typically replaced by a semiconductor and a non-linear (p-n, Schottky, or MOSFET) electronic device is then devised. This kind of device, while having a tremendous increase in sensitivity, becomes much more fragile and structurally unstable, especially in environments where high temperatures associated with corrosive or oxidizing gases are expected to be monitored. Recent literature [1] mentions Au, Ti, Ta, W as good choices for creating ohmic contacts on silicon carbide for hot environment applications. However, our previous work [2] show that Au and Ti contacts don't maintain structural integrity for extended time when operating at 800°C.

The work we present here deals with the evolution of the tungsten electrical contact layers deposited using e-beam evaporation from WC powder. We implanted and then heat treated (HT) our samples in air for 1 hr at 800 °C, using Rutherford Backscattering Spectrometry (RBS), Optical Absorption Spectrophotometry (OAS), and Raman Spectroscopy (RS) measurements to characterize the evolution of the deposit at the end of each preparation stage.

Experimental

Semi-insulating 6H silicon carbide was masked and deposited (e-beam evaporation from WC powder in a graphite crucible) with a thin (sub micron) tungsten carbide layer to be used as electrical contacts. We carried out RBS measurements using 2.1 MeV ^{4}He^{1+} ions in an IBM geometry with the particle detector placed at 170 deg. on both silicon carbide material and tungsten carbide deposits, RS on the tungsten deposits only, and OAS in the range 350 to 2000 nm on the silicon carbide chip only. Then, 1×10^{15} platinum ions/cm^2 with an energy of 13 keV (10±2 nm range) were implanted into the silicon carbide surface (and tungsten carbide contacts) at a 5 deg off-axis incident angle (to avoid channeling effects) to obtain a ~2 at% density of the platinum distribution in the penetration range. The set of characterization measurements were carried out again on the as implanted sample to observe the modifications brought by the ion implantation. Then the samples were annealed in air at 800 °C for 1 hr, followed with another complete round of characterization measurements.

Results and Discussions

RBS analysis using RUMP [3] show that the real composition of the layer deposited is much more complex, with a ratio W: of 3:997, and containing additional impurities associated with the operational history of the evaporation chamber (Bi, Te, Ag etc.). The total thickness of tungsten-containing layer was estimated to be 214 nm. The subsequent ion implantation had negligible effect on the layer's composition and elemental distributions. However, the HT in air for 1 hr seems to have had a major effect: most of the contaminants were driven away from the sample, as expected, while the tungsten distribution was negligibly affected by this process (see Figure 1).

Raman measurements carried out with a Jobin-Yvon confocal micro-Raman system using a He-Ne laser (632 nm wavelength) show a highly disorganized graphitic content in the deposited layer, that was largely un-affected by the ion implantation, and that disappeared after the heat treatment (see Figure 2). The small peaks that are visible on the heat-treated sample at 1380 and 1520 cm^{-1} belong to native graphitic regions from the bulk of the silicon carbide material. Also, the peaks in the 1600 to 1800 cm^{-1} region are characteristic to silicon carbide and become visible once the mostly graphitic layer on top has disappeared and the remaining tungsten layer was quasi-transparent to the He-Ne laser light.

Optical absorption measurements (Figure 3) show that there is significant surface roughness of the as deposited layer (overall 0.4 absorbance throughout the wavelength range) that decreases after the ion implantation (to ~0.2 absorbance). An interesting feature revealed after the implantation process is the shoulder formed at ~850 nm (1.46 eV) indicating that one of the impurities in the deposit are forming a highly organized macro-scale semiconducting structure with a bandgap of ~ 1.46 eV. As the impurities are driven away from the sample during the heart treatment, we notice that the signal due to this structure decreases in intensity, and we expect it to disappear completely during further heat treatment steps. There is an additional feature appearing on the as deposited and as implanted spectra, a small peak at ~625 nm, that also disappears after heat treatment. Since tungsten seems to still be present in the layer after heat treatment, we attribute this peak to one of the impurities, too.

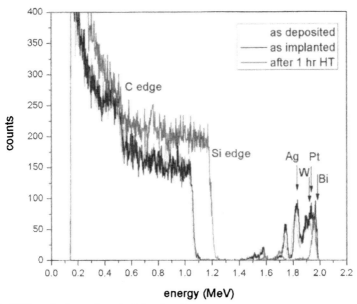

Figure 1. RBS on the layer deposited for electric contacts.

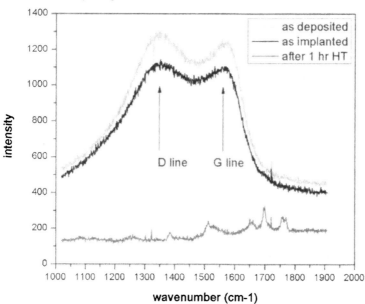

Figure 2. Raman spectra showing the D and G lines corresponding to highly distorted graphitic agglomerations in the investigated layer.

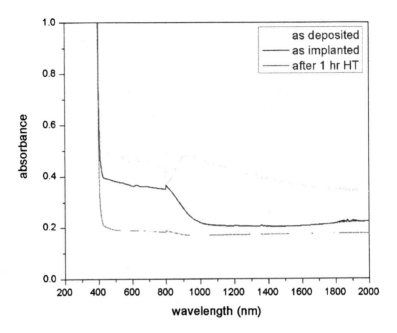

Figure 3. OAS data. The sudden jump at 800 nm is due to instrumental artifacts (filter changing system) and is not related to the sample itself.

Conclusions and future directions

Although the evaporation process proved to be faulty (we introduced large amounts of impurities to the system), we could drive most of them away from the layer during the heat treatment. The tungsten content of the layer has not changed as far as we could tell by RBS; however, its full characterization has been hampered by the presence of the additional contaminants. We also need to reduce the preferential deposition of carbon from the WC powder precursor, such that we could improve on the W:C ratio.

RBS data also shows that the integrity of the silicon carbide substrate is better preserved if we first deposit the electrical contacts and then we do the ion implantation and we are going to maintain this arrangement, rather that the opposite way, as we did in all previous work.

Overall, it seems that tungsten or tungsten carbide will work as a stable electrical contact on silicon carbide and once we will be able to obtain good tungsten carbide films (ratio in the single digits), we will pursue electrical characterization of such devices at 800°C.

Acknowledgements

This research sponsored by the Center for Irradiation of Materials at Alabama A&M University and the Center for Advanced Propulsion Materials at Alabama A&M University Research Institute, under contract number NAG8-1933 from NASA.

References

1. J. O. Olowolafe et al., Thin Solid Films 479 (2005) 59-63.
2. C. I. Muntele, S. Sarkisov, I. Muntele, D. ILA, Mat. Res. Soc. Symp. Proc. (2005).
3. L. R. Doolittle, M. O. Thompson, RUMP, Computer Graphics Service, 2002.

Mater. Res. Soc. Symp. Proc. Vol. 929 © 2006 Materials Research Society 0929-II03-04

SiC Based Neutron Flux Monitors for Very High Temperature Nuclear Reactors

Wolfgang Windl[1], Behrooz Khorsandi[2], Weiqi Luo[1], and Thomas E. Blue[2]
[1]Materials Science and Engineering, The Ohio State University, Columbus, OH, 43210
[2]Nuclear Engineering, The Ohio State University, Columbus, OH, 43210

ABSTRACT

The Gas Turbine-Modular Helium Reactor (GT-MHR) and the Very-High-Temperature Reactor (VHTR) are next-generation high-temperature reactor types that are being designed to operate under normal conditions with primary coolant outlet temperatures in the range of 850 °C and 1000 °C, respectively. A new type of silicon carbide based diode neutron detector is currently under development in order to monitor the neutron flux in this environment. An important problem, in this context, is the long-time reliability of the diodes under continuous irradiation at high temperatures. In this paper, we discuss a computational methodology to study the accumulation of radiation damage in the detectors as a function of temperature and its influence on the electrical properties.

INTRODUCTION

The Gas Turbine-Modular Helium Reactor (GT-MHR) is designed to operate with an outlet coolant temperature of approximately 850 °C. The Very-High-Temperature Reactor (VHTR) is being designed to operate with even higher outlet coolant temperatures. If a detector were to be placed within the GT-MHR or VHTR cores, it would be exposed to a very harsh environment, in the sense that the detector would be exposed to high temperatures and high neutron fluences. If neutron detectors are to be placed in environments that are as harsh as these, then novel detectors need to be developed to monitor the neutron flux. Because of its superior materials properties, we are investigating a hexagonal polytype of silicon carbide, 4H-SiC, as the base material for diode detectors. It is our hope that such detectors may be designed to survive long-time irradiation (at least one refueling cycle, i.e. 15.7 months for GT-MHR) without too much degradation in their electrical properties. This might allow them to be placed inside the reactor core, where they can be used to gather information regarding the neutron flux (and power) spatial distributions during reactor operation. With a more detailed knowledge of the neutron flux and power spatial distributions, engineering margins may be decreased, thus allowing for increases in the power level at which the reactor may be safely operated and thus reactor profitability.

In comparison to silicon (Si), SiC is a wide-band gap (3.2 eV for 4H-SiC) and radiation-hard semiconductor material. However, like Si, the properties of SiC change with irradiation by energetic neutrons, which cause displacement damage. As the neutrons collide with the atoms of the semiconductor, they may transfer enough energy to Primary Knock-on Atoms (PKAs) to displace them from their original sites and may, if sufficient energy is transferred to the PKA, initiate the creation of a damage cascade as the PKA collides with other atoms and moves them from their original lattice sites. As a result of the interactions described above, defects (such as vacancies, interstitials and antisites) are created. A significant fraction of the defects, which are initially formed, vanish as a consequence of recombination among the defects. However, a large enough fraction of the defects remains so that, with prolonged irradiation, the electronic characteristics of the detectors may change. These changes may alter the count rate of the detector, and hence may limit the lifetime of the detector. Accurately modeling the formation and evolution of the various types of defects, in SiC diode detectors as a consequence of neutron irradiation, is therefore an important step for accurately predicting the effects of neutron irradiation on the detectors and estimating the detectors' lifetime in various reactor environments.

In this paper, we describe our modeling methodology to approximate the number of defects initially created in thin SiC diode detectors by irradiation with monoenergetic neutrons with energies ranging from 50 eV to 7 MeV. The neutron energy spectra that we use are modeled spectra for selected locations in the central reflector of the GT-MHR. For the purposes of brevity and specificity, we focus our discussion of defect formation on C and Si vacancies.

COMPUTATIONAL METHODS

The SiC volume in our currently considered design for Schottky diode detectors is small, consisting of a 10 μm active layer on a 300 μm substrate. Consequently, the majority of neutrons pass through the SiC layer without interacting. Therefore, one can assume – especially when the detectors are first irradiated and the damage concentrations are low – that the damage cascades from different neutron impact events are well separated from one another.

We use the continuous energy Monte Carlo code MCNP5 [4] for modeling neutron trajectories in 4H-SiC. The neutron histories are tracked by the use of evaluated cross sections data libraries for elastic scattering, inelastic scattering and (n,gamma) absorption on the Si and C nuclei of the 4H-SiC structure. Besides obtaining the fraction of scattered neutrons from the output, we also can extract the characteristics of the scattered neutrons (energy, position and direction cosines), before and after each collision, as well as the type of the PKA they have interacted with, from the particle-track files created by MCNP5. The PKA energies and direction cosines are determined from the scattered neutron characteristics, using conservation of energy and momentum. The PKA type, position, energy and direction cosines serve as input for the TRIM [3] and MARLOWE [5] codes, which estimate the number of C- and Si-vacancies for C- and Si-PKAs.

Both TRIM and MARLOWE are based on the binary collision approximation, but they differ in their treatment of the target material. Whereas TRIM assumes that the target has an amorphous structure, MARLOWE allows the user to specify a crystalline target. Therefore, the damage determined by MARLOWE is direction dependent, and, as such, it allows the user to observe structure-dependent phenomena such as channeling. Marlowe also includes, in its output, predictions for the number of interstitials of various types that are created in addition to the number of vacancies and replacements and their positions in the target, which is supplied by TRIM as well.

Apart from that, both codes can determine the full damage cascade by allowing each PKA to recoil and cause its own sub-cascade of target damage. During the modeling of a damage cascade, an atom that has been hit is displaced, if the energy transferred to it is larger than a species-specific displacement threshold energy, E_d [3]. For our simulations, we use displacement-threshold energy values of 20 eV and 35 eV for C and Si, respectively, as suggested in Ref. [6].

Finally, we quantify the total damage created as a consequence of neutron irradiation by calculating the number of vacancies created per atom in the SiC diode (N_{VPA}) using

$$N_{\text{VPA}} = \frac{v\, N_{\text{PKA}}}{V\, n}, \tag{1}$$

where ν is the total number of vacancies created per PKA, N_{PKA} is the total number of created PKAs, V is the (irradiated) volume of the detector, and n is the atom density of Si and C atoms.

The major limitation to this (zero-temperature) approach is the absence of thermal damage annealing, which becomes especially important for high-temperature environments, such as the GT-MHR. Thus, the amount of damage calculated represents the upper limit that is due to the kinetic effects of the ion on the target atoms. In order to include damage annealing into the simulation, we are currently developing a kinetic lattice-Monte Carlo (KMC) code based on kinetic parameters from ab initio calculations, which will simulate the effect of finite temperature on the damage cascades created by TRIM and/or MARLOWE. By combining the KMC code with MCNP5 and TRIM/MARLOWE, we will be able to simulate continuous irradiation/annealing conditions and the long-term damage evolution in a high-temperature reactor environment.

As the last step of our work, we will relate the created damage to the electrical transport properties of the SiC n-type detector material. Following the approach of Ref. [7], the mobility μ, for an applied voltage in the α direction resulting in a current in the β direction, is given by

$$\mu_{\alpha\beta} = -e \sum_n \int \frac{d^2k}{(2\pi)^2} \tau_n(\mathbf{k}) v_\alpha^n(\mathbf{k}) v_\beta^n(\mathbf{k}) \frac{\partial f_0 E_n(\mathbf{k})}{\partial E}, \qquad (2)$$

where f_0 is the equilibrium Fermi distribution, $\mathbf{v}_n(\mathbf{k}) = \nabla_k E_n(\mathbf{k})/\hbar$ is the group velocity, and $\tau_n(\mathbf{k})$ is the lifetime of state $(n\mathbf{k})$. The lifetime can be computed from the scattering rates of the electrons using Fermi's Golden Rule. We calculated the necessary scattering matrix elements $\langle n\mathbf{k} | \Delta V | n'\mathbf{k}' \rangle$ from the difference in self-consistent potential between the structure with a defect and a defect-free reference structure (which can be, e.g., bulk Si or an interface structure), and use the effective-mass approximation for the conduction-band wave functions. The self-consistent potentials have been determined using the Vienna Ab-Initio Simulation Package (VASP) [8] with ultrasoft pseudopotentials within the generalized-gradient approximation [9].

For small enough concentrations of scatterers, the lifetime is inversely proportional to the scatterer density x and can be added to the lattice scattering mobility μ_L using Matthiessen's rule,

$$\mu_{tot}(x) = \frac{1}{1/\mu_L + x/\mu_x^0}, \qquad (3)$$

where μ_x^0 is a normalized scatterer-specific mobility, which allows one to compare the detrimental effects of different scatterers on the mobility. For different defect species, the denominator on the right-hand side can be extended by additional terms for the scattering from different impurities. The agreement of Eq. (3) with experiment can also be improved by allowing a constant mobility offset on the right-hand side.

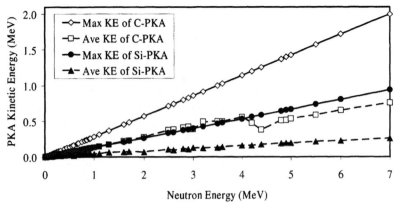

Figure 1. The maximum and average kinetic energies E_{kin} transferred to C- and Si-PKAs after collision with monoenergetic neutrons of varying energy.

RESULTS

Figure 1 shows our MCNP5 modeling results for the maximum and average C- and Si-PKA energies as a function of the neutron energy E_n. Theoretically, the ratio of the average PKA energy to the maximum PKA energy should be 0.5 for elastic and isotropic scattering. The ratio is ~0.5 for E_n up to ~100 keV for Si, and E_n up to ~1 MeV for C. Above these values, that ratio changes dramatically, particularly for Si-PKAs, indicating that the scattering is not isotropic anymore.

For elastic and isotropic scattering, the ratio of the average Si-PKA energy to the average C-PKA energy should be 0.469, theoretically. Our calculations show that the ratio of these two average energies indeed is 0.467 for low E_n, where collisions are elastic and isotropic. However, as E_n increases, this ratio changes significantly as well due to anisotropic scattering.

Figure 2 shows the results of the calculations for the number of vacancies per atom per fluence as a function of neutron energy, using TRIM as the binary-collision code. Interestingly, the

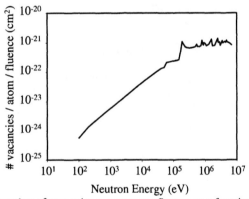

Figure 2. Calculated number of vacancies per atom per fluence as a function of neutron energy.

number of vacancies per atom per fluence is almost constant for neutron energies larger than 200 keV, because 1) at high neutron energies, the scattering angle for neutrons tends to be small, causing neutrons to transfer fractionally less energy to PKAs; 2) for neutron energies higher than 1.3 MeV, part of the transferred energy may excite PKAs, thus decreasing the PKA kinetic energy; 3) part of the PKA energy, especially at high PKA energies, goes into electronic excitations. Thus, we are able to model the damage creation process with well calibrated off-the-shelf codes. Once the kinetic Monte-Carlo modeling is in place and calibrated, we should be able to realistically model damage creation and evolution in 4H-SiC based detectors.

Concerning the modeling of the electron transport, to date we have applied our methodology only to the case of silicon, which is similar but simpler due to its cubic lattice structure. Figure 3 shows the normalized mobilities for a number of defects and impurities, calculated from self-consistent potentials determined from 64-atom supercell calculations. Figure 3 also shows the self-consistent scattering potentials for substitutional arsenic and vacancies. The scattering potential, in the case of arsenic, has very small values (dark blue/light gray) in most areas of the simulation cell and only has larger values in a small area around the arsenic atom (center of the cell). In the case of the vacancy, the scattering potential has large values in a significant portion of the simulation cell, explaining the predicted increased scattering probability of the electrons and thus the decreased mobility. Similar calculations for the case of 4H-SiC will allow us to relate the simulated damage to the electrical properties of the diode and thereby close the loop to predict the detector lifetime.

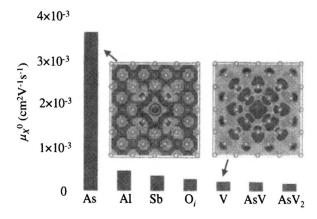

Figure 3. Normalized scattering mobilities for different impurities and defects in silicon. Shown are also the self-consistent scattering potentials, calculated within 64-atom supercells, for a substitutional As impurity and a vacancy.

CONCLUSIONS

4H-SiC diode detectors exhibit promising properties as neutron power monitors for high temperature nuclear reactors. Nevertheless, the electrical properties degradation of 4H-SiC, due to the harsh environment in the high temperature reactors, may restrict the use of this semiconductor material for in-core power monitors. In this paper, we focused on our modeling efforts to develop methods to predict the damage creation, recovery and the effects of defects on the electrical properties of 4H-SiC. In the near future, we will match these results with experimental data to predict the lifetime of 4H-SiC semiconductors in high temperature reactors.

ACKNOWLEDGMENTS

This material is based upon work supported by the US Department of Energy under the NERI program Award No. DE-FG-07-02SF22620 and NERI Project Number 02-207. Any opinions, findings, and conclusions or recommendations expressed in this material are those of the authors and do not necessarily reflect the views of the Department of Energy. WW also acknowledges funding from the Semiconductor Research Corporation under contract number 2002-MJ-1018 and the National Science Foundation under contract number 0244724.

REFERENCES

[1] L.R. Greenwood, and R.K. Smither, "SPECTER: Neutron Damage Calculations for Materials Irradiations," ANL/FPP/TM-197, Jan. 1985.

[2] M.B. Lee, and E.H., Farnum, "The Effect of Neutron Energy on Defect Production in Alumina," *Nuclear instruments and Methods in Physics Research,* Vol. B, No. 102, 1995, pp. 113-118.

[3] F.J. Ziegler, "SRIM-2003," *Nuclear Instruments and Methods in Physics Research,* Vol. B, No. 219-220, 2004, pp. 1027-1036.

[4] R.A. Forster, et al "MCNP Version 5," *Nuclear Instruments and Methods in Physics Research,* Vol. B, No. 213, 2004, pp. 82-86.

[5] M. T. Robinson, *MARLOWE: Binary Collision Cascade Simulation Program, Version 15b, A Guide for Users* (December 5, 2002), http://www-rsicc.ornl.gov/codes/psr/psr1/psr-137.html.

[6] R. Devanathan, W. J. Weber and F. Gao, *Atomic scale simulation of defect production in irradiated 3C-SiC,* J. Appl. Phys. **90**, 2303 (2001).

[7] M. H. Evans, X.-G. Zhang, J. D. Joannopoulos, and S. T. Pantelides, *First-Principles Mobility Calculations and Atomic-Scale Interface Roughness in Nanoscale Structures,* Phys. Rev. Lett. **95**, 106802 (2005).

[8] G. Kresse and J. Hafner, Phys. Rev. B 47, 558 (1993); 49, 14251 (1994); G. Kresse and J. Furthmüller, Comput. Mater. Sci. **6**, 15 (1996); Phys. Rev. B **55**, 11169 (1996).

[9] G. Kresse and J. Hafner, J. Phys.: Condes. Matt. **6**, 8245 (1994).

Mater. Res. Soc. Symp. Proc. Vol. 929 © 2006 Materials Research Society   0929-II06-05

Compression Testing and Microstructure of Heat-Treatable Aluminum Periodic Cellular Metal

B. A. Bouwhuis, and G. D. Hibbard
Materials Science and Engineering, University of Toronto, 184 College Street, Toronto, Ontario, M5S3E4, Canada

ABSTRACT

Periodic cellular metals (PCMs) can offer higher specific strengths and stiffnesses than conventional (i.e. stochastic) metallic foams. This study examines the effects of PCM microstructure and loading conditions on the mechanical performance.

PCM cores with 95% open porosity were constructed from perforated 6061 aluminium alloy sheets using a perforation-stretching method. This method places planar, periodically-perforated sheet metal in an alternating-pin jig. The pins apply force out-of-plane, plastically deforming the sheet metal into a truss-like array of struts (i.e. metal supports) and nodal peaks (i.e. strut intersections). Micro-hardness profiles were taken in the PCM struts to investigate microstructural evolution during fabrication and after heat treatment.

Truss cores were tested in two limiting uniaxial compression conditions. In the first, the PCM cores are placed between smooth compression platens where the nodes are laterally free and compressive forces are resisted through PCM node-bending (i.e. free compression). In the second, the PCM cores were placed between plates where the nodes are laterally confined and compressive forces are resisted through PCM beam-buckling (i.e. confined compression). Compression response was analyzed in terms of peak compressive strength, elastic modulus, and energy density absorbed upon densification; response values were used to illustrate the effect of compression test conditions. In addition, PCM cores were tested in the age-hardened state and annealed state to determine microstructural effects on compressive response.

Analysis of PCM response in free- and confined-compression conditions indicates a greater force resistance in beam-buckling over node-bending resistance mechanisms. The compressive strength, elastic modulus, and energy density of heat-treatable AA6061 PCMs are be found to respond: 1) over a wide range of value, dependent on the microstructure; 2) over a wide range of value, dependent on the PCM compression conditions; and 3) equally, if not more repeatable and with higher compressive strength-to-weight ratio than conventional metal foams.

INTRODUCTION

Cellular materials have intrinsic properties that uniquely combine low weight with structural, vibration, acoustic, thermal, electrical, and catalytic function to offer great potential within many environments [1]. The Apollo command module was one such example; cellular honeycomb-core sandwich panels were used to construct a spacecraft that was light in weight but could also sustain the extreme temperatures and stresses of acceleration and landing [2]. More recently, periodic cellular metals (PCMs) have received considerable attention [3-6]. Open-cell PCM architectures [e.g. 7], reduce the total material mass by retaining only that which has geometrically-high load-bearing efficiency. The resultant strength-to-weight ratio is improved and relative density (i.e. density of the PCM divided by that of the parent material) can be reduced to as low as 2-3% [e.g. 7-9]. Furthermore, these architectures can be used as cores in sandwich structures and provide multi-functionality such as cross-flow heat exchange [10] and shape morphing [11]. Overall, PCM cores and sandwich structures may exhibit superior mechanical properties when compared to their stochastic metal foam counterparts [7].

The current research investigates the correlation between PCM fabrication, resultant microstructure, loading condition and compressive properties of heat-treatable aluminum alloy PCMs.

EXPERIMENTAL DETAILS

Pyramidal stand-alone PCMs were fabricated from a 0.81 mm thick (t) square punched aluminium 6061-T6 sheet, purchased from Woven Metal Products, Inc. (Alvin, TX). The 90.82 mm^2 (internal side length l = 9.53 mm) perforations were punched on a 2D square lattice of unit cell size 12.7 mm × 12.7 mm. The resulting structure is a series of 4-rayed nodes with arm cross-sections of 3.18 mm (w) × 0.81 mm (t), having 56% open area. The sheet was heat treated at 500°C for 30 minutes, followed by furnace cool to form a supersaturated solid solution [9]. Microhardness verified a reduction in hardness to the pre-fabricated condition of 41.4 ± 0.8 HV (50 gf load over 10 measurements).

PCMs can be fabricated using simple sheet forming operations [5-7]. For this experiment, a perforation-stretching process was used similar to that described by Sypeck and Wadley [12]. The linear deformation rate was ~5mm/min and a final pyramidal height of 7.3 mm was achieved. The result is a truss angle ω [e.g. 13] of ~30° and relative density of ~5% (0.13 Mg/m^3), PCM array shown in Figure 1a.

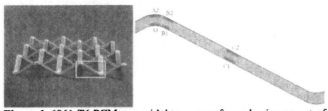

Figure 1. 6061-T6 PCM pyramidal truss core formed using an out-of-plane deformation method [e.g. 12]. PCM has truss angle ω = 30° and relative density of 5% (0.13 Mg/m^3). Right: a Vickers hardness profile through a PCM strut highlighting six regions of interest.

After forming, the struts had cross-section width = 3.08 ± 0.02 mm and thickness = 0.70 ± 0.02 mm over seven measurements. The PCM cores were cut into 36-node squares for an effective compression-surface area of 44.4 cm^2. To age-harden the PCM core, a treatment of 165°C for 19 hours was used to achieve the maximum (T6) hardness [9, 14].

Samples for microhardness testing were cast in epoxy and prepared using standard metallographic techniques [14, 15]. Microhardness measurements were performed with a Vickers indenter under a 50gf load and dwell time of 10 s, after ASTM-E 384 [16].

Uniaxial compression testing was performed at a crosshead displacement rate of 1 mm/min. Five test samples were used for each compression test condition, after ASTM-C365 [17]. Strains were measured from crosshead displacement [1, 18-22]. Mechanical response of PCM stand-alone structures has shown dependence on the testing conditions [23]; following this study, PCMs are tested in two conditions: 1) free and 2) confined compression.

In the free-PCM uniaxial compression mode, a single layer of pyramidal AA6061 PCM core is placed between tempered steel compression platens. This compression method illustrates force resistance by plastic bending (i.e. hinging) at the PCM nodes. The point of plastic hinging is linearly dependent on the yield strength of the material at the hinge location [1]; to investigate this relation, the hinge location is denoted by B1 and B2 in Figure 1b.

In the confined-PCM uniaxial compression mode, a single layer of pyramidal PCM core is fixed within two confinement plates [23] and placed between the compression test platens. These plates laterally confine the PCM nodes through the duration of compression. This method illustrates force resistance by an additional mechanism: plastic buckling of the PCM struts. The point of buckling is linearly dependent on the yield strength of the strut material [1, 4]; to investigate this relation, the strut location is denoted by C1 and C2 in Figure 1b.

The peak compressive strength, σ_P, of the PCM was taken as the initial peak stress [7, 9, 23]. Measurements of an elastic modulus E are made from the largest slope of the loading curve. In addition, an unloading/loading curve is also used to calculate $E_{75\%}$, the modulus unloaded from ~75% of σ_P, the same manner as used for conventional metallic foams [1, 22].

Measured energy absorption, J_D, during truss collapse is taken as the area under the σ-ε curve between strains at ½ σ_P and 2 σ_P [23, 24]; this area is calculated as an upper-bound integral using a Riemann sum [e.g. 25] with $\Delta\varepsilon \sim 10^{-4}$.

RESULTS/DISCUSSION

Figure 2 presents the microhardness values for regions A, B, and C of the as-fabricated (i.e. work-hardened), solutionized and age-hardened PCMs prior to compression. These values are in accordance with the expected hardness for AA6061 in the solutionized (30-40 HV) and T6 (100-110 HV) conditions [14]. For each PCM, Figure 3 displays the free and confined compression results, shown as a uniaxial compression stress-strain (σ-ε) curve, for all 5 test samples.

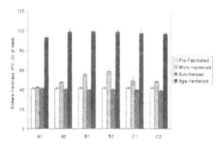

Figure 2. Microhardness measurements of regions A, B, and C (see Figure 1) for each PCM: as-fabricated (i.e. work-hardened), solutionized and age-hardened.

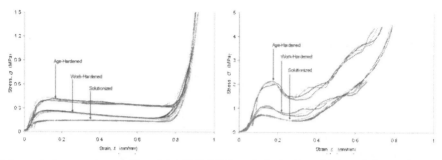

Figure 3. Compression profiles of AA6061 PCMs: solutionized, as-fabricated (i.e. work hardened) and age-hardened PCMs in both free (left) and confined (right) compression.

There is a gradual increase in σ_P from solutionized to work-hardened to age-hardened microstructure for both free and confined compression conditions (see Figure 3). Peak strength (in MPa) increases from solutionized (0.15 ± 0.01; 0.74 ± 0.03) to work-hardened (0.26 ± 0.01; 1.08 ± 0.03) to age-hardened (0.41 ± 0.02; 2.06 ± 0.06) for the free and confined compression condition, respectively.

The effect of microstructure on energy density to densification (i.e. J_D) is also apparent in Figure 3. A gradual increase in J_D (in kJ/m^3) occurs from solutionized (114 ± 6; 342 ± 16) to work-hardened (175 ± 5; 679 ± 28) to age-hardened (318 ± 13; 917 ± 22) PCMs for free and confined compression conditions, respectively. Between conditions, J_D shows a 200-340% increase in confined compression over free compression.

Measurement of modulus E is more difficult due to the initial bedding in of the PCM [9]. Ideally, PCM modulus is only dependent on the PCM architecture and the stiffness of the parent material [4, 13]. For free and confined compression, respectively, the increasing E (in MPa) from solutionized (2.5 ± 0.5; 12.4 ± 1.3) to work-hardened (5.3 ± 0.5; 18.9 ± 1.7) to age-hardened (6.4 ± 0.9; 33.2 ± 2.4) microstructures is likely related to the initiation of limited plastic deformation before the peak stress is achieved. Therefore, a measurement by unloading/loading at 75% of σ_P (i.e. $E_{75\%}$) was also used and resulted in the same free compression stiffness values for solutionized (8.2 ± 0.5 MPa) and age-hardened (8.1 ± 0.7 MPa) PCMs.

PCMs offer lower sample-to-sample variability in compression test behaviour than what is typically observed for conventional metallic foams. For example, the standard deviation in foam compressive strength is typically between 5% and 15% [1, 18, 22, 24]. Due to their periodicity, PCMs are postulated to have lower standard deviations than their stochastic foam counterparts [7]. The mechanical property data described supports this assertion, where standard deviation in compressive strength for all PCMs is below 5%.

The results obtained in Figure 3 illustrate the important point that both microstructure and testing conditions have a significant effect on the measured compression properties of identical PCM architectures. In this study, the solutionized PCMs can be used as a reference point for evaluating the effectiveness of work-hardening and age-hardening on the overall PCM peak compressive strength. Table I presents the normalized peak strengths for all three microstructures and from both loading conditions. Also shown in Table I are the normalized microhardness values for regions B (plastic hinge) and C (mid-strut). Several important observations can be made. First, there is approximately the same peak strength increase after age-hardening for both bending dominated (i.e. free compression, 2.75 ± 0.23) and buckling dominated (i.e. confined compression, 2.79 ± 0.18) failure mechanisms. This is consistent with the essentially uniform hardness increase provided by this strengthening mechanism (2.50 ± 0.11 in region B and 2.49 ± 0.11 in region C). On the other hand, work-hardening increases the strength in the hinge region (1.45 ± 0.12, region B) more than in the mid-strut region (1.25 ± 0.08, region C). Accordingly there is a proportionately greater strength increase during bending-dominated failure, i.e. when failure is controlled by the material strength at the plastic hinge of region B, than in buckling-dominated failure, i.e. when failure is controlled by the material strength at mid-strut (region C).

Table I. Relative magnitudes of peak strength σ_P for free and confined compression as well as microhardness for regions B (i.e. HV_B) and regions C (i.e. HV_C) (see Figure 1). Values are normalized with respect to solutionized PCM.

	Solutionized (reference)	Age-Hardened PCM	Work-Hardened PCM
σ_P, free	1	2.75 ± 0.23	1.73 ± 0.10
σ_P, confined	1	2.79 ± 0.18	1.46 ± 0.09
HV_B	1	2.50 ± 0.11	1.45 ± 0.12
HV_C	1	2.49 ± 0.11	1.25 ± 0.08

The present data can be used to illustrate the different strengths between node-bending (free-PCM) and strut-buckling (confined-PCM) failure mechanisms; conditions of confined and free compression can produce mixtures of these failure mechanisms. In practice, any internal joint of a PCM core behaves in a fashion intermediate to bending and buckling conditions [12]. Depending on the test conditions, the periodic framework in this study is not necessarily rigid, but instead can collapse through a range of bending and buckling responses. The present study serves as an initial step for microstructural engineering of the PCM compressive performance.

CONCLUSIONS

This study has investigated the deformation mechanisms of identical PCM structures in a range of compression testing conditions in two limiting microstructures. In free compression, the performance is determined by the bending strengths of the PCM nodes. In confined compression, the performance is determined by the buckling strengths of the PCM struts and successive weakening and strengthening mechanisms during truss collapse. By designing the fabrication process to produce a specific PCM microstructure, additional parameters become available to control the mechanical performance.

ACKNOWLEDGEMENTS

Financial support from the Natural Sciences and Engineering Research Council of Canada (NSERC) is gratefully acknowledged. One of the authors (B.A.B.) is supported by a University of Toronto Open Fellowship.

REFERENCES

1. M. F. Ashby, A. G. Evans, N. A. Fleck, L. J. Gibson, J. W. Hutchinson and H. N. G. Wadley, *Metal Foams: A Design Guide* (Butterworth-Heinemann, Boston, 2000) pp. 3-72.
2. J. M. Davies, *Lightweight Sandwich Construction* (Blackwell Science, Toronto, 2001) pp. 1.
3. M. F. Ashby and Y. J. M. Bréchet, *Acta Mater.* **51**, 5801 (2003).
4. M. F. Ashby, *Phil. Mag.* **85**, 3235 (2005).
5. D. J. Sypeck, *App. Compos. Mat.* **12**, 229 (2005).
6. H. N. G. Wadley, *Phil. Trans. Roy. Soc. A* **364**, 31 (2006).
7. H. N. G. Wadley, N. A. Fleck and A. G. Evans, *Compos. Sci. Tech.* **63**, 2331 (2003).
8. S. Chiras, D. R. Mumm, A. G. Evans, N. Wicks, J. W. Hutchinson, K. Dharmasena, H. N. G. Wadley and S. Fichter, *Int. J. Sol. Struct.* **39**, 4093 (2002).
9. G. W. Kooistra, V. S. Deshpande and H. N. G. Wadley, *Acta Mater.* **52**, 4229 (2004).
10. J. Tian, T. Kim, T. J. Lu, H. P. Hodson, D. T. Queheillalt, D. J. Sypeck and H. N. G. Wadley, *Int. J. Heat Mass Trans.* **47**, 3171 (2004).
11. D. M. Elzey, A. Y. N. Sofla and H. N. G. Wadley, *Int. J. Sol. Struct.* **42**, 1943 (2005).
12. D. J. Sypeck and H. N. G. Wadley, *Adv. Eng. Mat.* **4**, 759 (2002).
13. V. S. Deshpande and N. A. Fleck, *Int. J. Sol. Struct.* **38**, 6275 (2001).
14. J. R. Davis, *ASM Specialty Handbook: Aluminum and Aluminum Alloys* (ASM International, 1993) pp. 485-492, 686.
15. ASTM Standard E 3-99, *Standard Test Method for Preparation of Metallographic Specimens* (American Society for Testing and Materials, 1999).
16. ASTM Standard E 384-99, *Standard Test Method for Microindentation Hardness of Materials* (American Society for Testing and Materials, 1999).
17. ASTM Standard C 365, *Standard Test Method for Flatwise Compressive Properties of Sandwich Cores* (American Society for Testing and Materials, 1999).
18. A. E. Simone and L. J. Gibson, *Acta Mater.* **46**, 3109 (1998).
19. B. Krizst, B. Foroughi, K. Faure and H. P. Degischer, *Mat. Sci. Tech.* **16**, 792 (2000).
20. E. W. Andrews, G. Gioux, P. Onck and L. J. Gibson, *Int. J. Mech. Sci.* **43**, 701 (2001).
21. U. Ramamurty and A. Paul, *Acta Mater.* **52**, 869 (2004).
22. E. W. Andrews, W. Sanders and L. J. Gibson, *Mat. Sci. Eng. A* **270**, 113 (1999).
23. B. A. Bouwhuis and G. D. Hibbard, submitted to *Met. Mat. Trans. A* (2006).
24. O. B. Olurin, N. A. Fleck and M. F. Ashby, *Mat. Sci. Eng. A* **291**, 136 (2000).
25. S. Salas, E. Hille and G. J. Etgen, *Calculus: One and Several Variables*, 8th ed. (John Wiley & Sons, Inc., Toronto, 1999) pp. 264-266.

Mater. Res. Soc. Symp. Proc. Vol. 929 © 2006 Materials Research Society   0929-II06-06

Effects of Extreme Radiation Environment on Composite Materials

Jianren Zhou[1], Jerrel Moore[1], Vernon Calvin[1], Richard Wilkins[1], Sofia Martinez Vilarino[1], Yang Zhong[1], Brad Gersey[1], and Sheila Thibeault[2]

[1]College of Engineering, Prairie View A&M University, Prairie View, TX, 77446
[2]NASA Langley Research Center, Hampton, VA, 23681

ABSTRACT

Future manned space travel will go beyond the Lower Earth Orbit (LEO) into deep space exploration and providing protection from space radiation is a major challenge. It is essential to study the effects of the space environment on materials to ensure safe and successful missions.

This paper summarized the studies of two materials, *in-site* regolith composites and LTM 45 composites, for potential applications in space radiation environments. The effects of radiation on the mechanical and thermal properties of the composites were investigated. The radiation shielding effectiveness of *in-situ* composites and low temperature molding materials was analyzed. The work was part of the efforts in studying and development of the multifunctional materials for long-term radiation exposures, to ensure effective radiation shielding and maintaining integrity of materials' mechanical and thermal properties for future space missions.

1. INTRODUCTION

The ionizing radiations in space affecting human operations are of three distinct sources and consist of a variety of energetic particles including ions formed by stripping the electrons from all of the natural elements. The three sources of radiations are associated with different origins identified as those of galactic origin (galactic cosmic rays, GCR), particles produced by the acceleration of solar plasma by strong electromotive forces in the solar surface and acceleration across the transition shock boundary of propagating coronal mass ejecta (solar particles event radiation, SPE), and particles trapped within the confines of the geomagnetic field [1, 2]. SPEs are associated with solar flares which produce intense burst of high energy plasma propagating into the solar system. The two sources of radiation that are especially hazardous for deep space exploration missions are SPEs and GCR.

Not only will the crew be exposed to space radiation during the trips in space but also in Moon or Mar's surface. Mars, unlike the Earth, lacks an intrinsic magnetic field and has a much thinner atmosphere. Therefore, the surface of Mars receives much more radiation from GCR and SPE than the Earth. Human explorers, as well as electronic devices, must be effectively protected from the radiation [1-2].

The high costs of added radiation shielding are a potential limiting factor in deep space missions [4]. Multifunctional materials (for example, structural elements which have good shielding properties) will be common in the optimization process. Such materials will provide adequate radiation shielding, required mechanical and thermal properties, and ease of processing.

The use of *in-situ* lunar/Martian materials to build shields for inhabitants is not only of convenience but also needed to decrease launch weight, which, in turn, will significantly lower the mission costs. Martian regolith (surface dirt) has been evaluated to be an economic resource for this purpose [5-6]. In this study polyimide was used as a binder for the regolith composites. In addition, low temperature molding composites LTM45 were also studied due to their reduced manufacturing costs and enhanced mechanical performance [7].

Polyimide bonded regolith panels were exposed to the radiation of proton and neutron to simulate the space radiation conditions. Radiation shielding effectiveness was characterized by measuring the fluence and the errors occurred to electronic chips behind the regolith panels. Compression tests and thermal analysis were performed on the regolith specimens before and after irradiation to study the effects on the mechanical and thermal properties of the candidate habitat materials [8].

Low temperature molding composites were exposed to two different types of radiation, a high-energy neutron (1-800 MeV) beam and a mid-energy proton (40 MeV) beam [7]. After irradiation, the mechanical properties of the composites were tested to understand the effects of the exposure on the materials.

2. EXPERIMENTAL

2.1 Radiation Sources

The ground based radiation sources used in these experiment were chosen to be relevant to various space radiation environments. Protons with energies in the tens of MeV are prevalent in the LEO environments [2] The energies for these proton experiments described here were chosen because of beam availability (Texas A&M Cyclotron) and because they represent significant components of the proton spectra in LEO orbits [2]. The 800 MeV protons (Los Alamos Neutron Science Center, LANSCE) used for some of the experiments are relevant to the GCR spectrum [1]. The broad spectrum (1-800 MeV) neutron source (LANSCE) has been shown to be similar to the secondary neutron spectra measured onboard spacecrafts like the MIR and International Space Station (ISS) [9]. These secondary neutrons are the result of GCR interactions with the spacecraft materials and can be very biologically damaging [1].

2.2 Materials

Polyimide bonded regolith panels were provided by NASA Langley Research Center (LARC). The polyimide used was LaRC-SI, a soluble imide developed by LARC. The formulation of LaRC-SI and the chemical composition of the regolith can be found in [10]. Two different regolith/polyimide ratios were provided. The first one, referred to as Regolith 101, contained 40% LaRC-SI with a thickness of 1.89 g/cm^2, and the second one, Regolith 102, 20% LaRC-SI with a thickness of 2.01 g/cm^2.

Specimens of 15.24 x 15.24 cm were cut for radiation exposure tests in which the fluence behind the shielding regolith material was measured. Then the panels were cut into compression testing specimens with dimensions about 1.27 x 1.27 x 2.54 cm.

LTM45 is a low temperature curing epoxy resin with improved toughness. The LTM45 composites were fabricated at Prairie View A&M University. First, the prepreg sheets were cut into 30.48 cm squares to produce a 16 layer laminated plate. Each 16-layer plate was mounted in a vacuum bag and fully cured on a Wabash press. The press-heating rate was two degrees per minute up to a temperature of 177 °C, and then the plate was held for 30 minutes at a pressure of 689.5 kPa.

After curing, the laminated plates were cut into ASTM 790 standard size, 2.54 cm x 6.35 cm, for mechanical flexure tests. The flexure test specimens were cut with the fibers aligned in the longitudinal direction and transverse direction, i.e., 0-degree ([0]$_{16}$) and 90-degree ([90]$_{16}$), respectively. The specimens had a nominal thickness of 0.254 cm. Four specimens were cut into 10.16 cm x 10.16 cm for shielding tests.

2.3 Radiation Tests

The proton exposures of the regolith based materials and the LTM45 composites were performed by the K500 superconducting cyclotron at Texas A&M Cyclotron Institute. The polyimide bonded regolith was irradiated with a 55 MeV proton beam while the LTM45 composites were exposed to a proton beam of 40 MeV.

Two types of tests were performed during the exposures for the regolith materials. The first was to measure the fluence behind the shielding regolith panels, in which a 15.24 x 15.24 cm specimen was placed in front of a 2.54 cm diameter beam and irradiated with various levels of fluence and flux. A detector was placed on the other side of the regolith panel to measure the proton fluence behind the specimen.

The other exposure tests performed on the regolith materials were irradiation on the specimens which would be then undergone mechanical testing. The regolith panels were first cut into compression specimens with dimension of about 1.27 x 1.27 x 2.54 cm and then exposed to the irradiation.

The regolith composites and LTM45 composites were irradiated by neutron at the Weapon Neutron Research facility at Los Alamos Neutron Science Center with the broad spectrum 1-800 MeV neutron beam. Both shielding effectiveness measurements and irradiation on regolith specimens for mechanical testing were performed. The set-up of the shielding effectiveness tests for neutron exposure, with a beam diameter of 7.62 cm, was similar to that in proton exposure. Instead of using a fluence detector behind the regolith panel, a microelectronic test bed with 8 lined-up integrated circuit (IC) chips [11] was placed behind the sample panel and the number of error occurred in those chips was recorded. These experiments were designed to test the shielding effectiveness for electronic devices by measuring the effect of shielding materials on a selected figure of merit from the device. In the case of these ICs, single event upsets (bit flips) in memory cells were counted as a function of independently measured radiation fluence. The 800 MeV proton exposures on regolith based materials were also performed at Los Alamos Neutron Science Center in which only samples for mechanical testing were exposed.

2.4 Mechanical Testing

The Instron mechanical testing machine, model 5582, was used to perform mechanical compression tests on the specimens with a crosshead speed of 0.127 cm/min. Baseline regolith samples used for mechanical compression testing were cut from the 15.24 x 15.24 cm panels into specimen sizes according to ASTM D695.

For the LTM45 composite materials, stress and modulus were determined by following ASTM 790 standard for a three point flexural test and using the Instron machine.

2.5 Thermal Analysis

The glass transition temperature, T_g, of the regolith materials was determined using a Differential Scanning Calorimeter machine, model 2910 by TA Instruments. The specimens were placed in aluminum pans, mechanically sealed and heated from room temperature to 275°C at 5°C/min.

Glass transition temperature was measured for LTM45 composites by DMA, model 2980 by TA Instruments. The samples were cut into rectangular specimens and run from room temperature to 300°C at a rate of 5°C/min.

3. RESULTS AND DISCUSSIONS

3.1 Proton Radiation

The shielding effectiveness of the regolith materials was characterized using 55 MeV proton exposures and measuring the fluence passing through the material panels. When the flux was varied from 1×10^5 to 1×10^7 protons/cm^2/s and, correspondingly, the fluence from 1.0×10^6 to 4.0×10^9 protons/cm^2, the percentage of fluence shielded by the regolith material was calculated. The values were found to be a constant with samples containing 60% regolith shielding off 19% radiation while samples containing 80% regolith shielding off 33% under all these flux and fluence conditions. The shielding effectiveness under other incident fluences needs further study.

In order to understand the effect of regolith content on the shielding effectiveness of the polyimide bonded regolith materials, the percentage of fluence shielded by the material was plotted versus regolith content, as shown in Figure 1. It can be seen that the addition of 20% more regolith, from 60% to 80%, could increase the shielding effectiveness. Comparing this result with the annual dose equivalent data reported by Kim et al [6], the data indicates that the regolith has more shielding effectiveness to medium-energy proton while polyimide resin is more effective to the complete galactic cosmic spectrum.

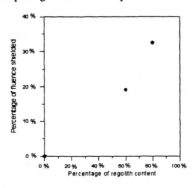

Figure 1. Percentage of fluence shielded by the material vs. regolith content.

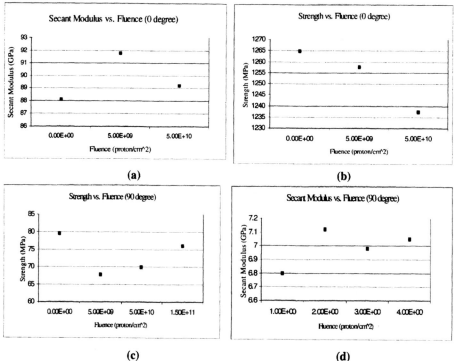

Figure 2. Effects of the mid-energy proton radiation on strength and modulus in flexural tests on LTM45 composites: (a) & (b), 0°; (c) & (d), 90°.

Figure 2, from (a) to (d), shows scatter plots based on flexural test data of LTM45 specimens irradiated with the 40 MeV proton beam. These are the averages of 90-degree and 0-degree specimens. In addition, a group of 90-degree and 0-degree specimens were pre-stressed before they were irradiated with the mid-energy proton beam. It can be seen that the strength of 0-degree specimens showed a decreasing trend with an increase in proton irradiation duration, while the secant modulus of 90-degree specimens had a slight increase with increasing proton fluence.

3.2 Neutron Radiation

The shielding effectiveness of the regolith materials to neutron radiation was characterized by detecting the single event upsets (SEU or bit flip errors) on a microelectronic testing bed [11] placed behind the regolith panel. The fluence used for all of the tests was 5.0×10^9 neutrons/cm^2 with a flux of 3.4×10^4 neutron/cm^2/pulse at 90 pulse/s. The number of errors occurred on each chip and the total number of errors are listed in Table I.

The results showed that both specimen panels, containing 60% and 80% regolith, perhaps caused a slightly increase of errors in the electronic board, as compared with the situation without any shield. However, it did not result in any strong secondary radiation that would cause substantially more errors. Therefore, further experiments with thicker panels should be conducted in order to evaluate the influence of the thickness on the shielding effectiveness against neutron radiation.

For the LTM composites, the shielding effectiveness under neutron irradiation was measured by a Tissue Equivalent Proportional Counter (TEPC) in terms of absorbed dose. The absorbed dose actually measures the energy that is deposited in some given mass. The shielding effectiveness testing showed the absorbed dose to tissue decreased with addition of the composites panels. At a density thickness of 40 g/cm^2, the absorbed dose to tissue was reduced by about 2/3 as compared to the blank test, as shown in Figure 3.

The SEU experiments and TEPC measurements were attempts to measure the shielding effectiveness of these materials for both spacecraft electronics (SEU) and humans (TEPC) in relevant environments separately.

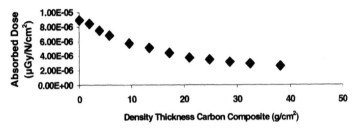

Figure 3. Absorbed dose vs. density thickness for LTM45 composite.

3.3 Mechanical Properties

One of the main concerns when a composite material is exposed to long-term radiation is the effect that the radiation may have on the molecular structure of the binder polymer [10]. There is a chance that the chains of the polymer may break, lowering the strength of the bulk structural material. In this research work the polyimide bonded regolith specimens were tested for compression properties prior and after various radiation exposures to investigate radiation effects on the potential structural shielding material.

The test results of compression strength and modulus of the regolith material after radiation exposures under 55 MeV proton, neutron, and 800 MeV proton are listed in Table II. The data showed that the compression strength and modulus of material containing 80% regolith were 20% and 36% higher than those with 60% regolith, respectively.

Both sample materials, Regolith 101 and 102, did not show conspicuous changes in the compression strength under the radiation conditions used in this work. However, the modulus of the Regolith 101 dropped about 16% after being irradiated by 800 MeV proton with a fluence of 5.34×10^9 proton/cm^2. With an even higher fluence of 1.78×10^{10} proton/cm^2, the modulus of the 60% regolith system decreased by about 20% from the baseline value. The drop in modulus of the polyimide bonded regolith after the high energy proton exposure indicated that the radiation did cause some damages to the molecular structures of the binder polymer. Further studies are needed to identify damages at molecular scales. The damages are believed to affect mainly the polymer, therefore it is reasonable to think that when the regolith content was increased the drop on the mechanical properties would be smaller after irradiation. This can be observed for Regolith 102 (with 80% regolith content) in which the decrease of strength after 800 MeV proton radiation with a fluence of 1.78×10^{10} protons/cm^2 was 6% and with the neutron exposure, the drop was almost the same as that for the baseline.

Table I. Shielding effectiveness of polyimide bonded regolith on electronic chips under neutron radiation

Shielding materials	Chip No.	Beam focused on chip 1, 2, 4		Beam focused on chip 20, 40, 80	
		Errors on each chip	Total errors	Errors on each chip	Total errors
Baseline (No shield)	Chip 1	330		0	
	Chip 2	64		0	
	Chip 4	71		0	
	Chip 8	8		0	
	Chip 10	0		7	
	Chip 20	1	474	257	421
	Chip 40	0		89	
	Chip 80	0		68	
Regolith 101 (Polyimide/regolith 40/60)	Chip 1	369		0	
	Chip 2	105		0	
	Chip 4	95		0	
	Chip 8	8		2	
	Chip 10	1		8	
	Chip 20	1	580	275	483
	Chip 40	0		128	
	Chip 80	1		70	
Regolith 102 (Polyimide/regolith 20/80)	Chip 1	356		0	
	Chip 2	179		0	
	Chip 4	88		0	
	Chip 8	8		0	
	Chip 10	0	633	7	
	Chip 20	1		277	428
	Chip 40	1		76	
	Chip 80	0		68	

Table II. Compression properties of polyimide bonded regolith before and after radiation exposures

Radiation conditions	Regolith 101 Polyimide/regolith 40/60		Regolith 102 Polyimide/regolith 20/80	
	Strength (MPa)	Modulus (GPa)	Strength (MPa)	Modulus (GPa)
Baseline (No radiation exposure)	193.23	5.93	231.52	8.07
55 MeV proton (Fluence: 1×10^6 protons/cm^2)	190.99	5.88	243.86	9.11
800 MeV proton (Fluence: 5.34×10^9 proton/cm^2)	187.66	4.96	235.52	7.94
800 MeV proton (Fluence: 1.78×10^{10} proton/cm^2)	180.10	4.81	218.70	7.94
Neutron (Fluence: 5.0×10^9 neutron/cm^2)	187.27	5.24	231.65	7.80

The effects of neutron radiation on the flexural properties of LTM45 composites are summarized in Table III. In addition, Figure 4 (a)-(d) shows the variation of material's flexural strength and modulus as a function of fluence of neutron irradiation. Five samples of both 0 and 90-degree specimens were tested for each radiation condition, and the results on Figure 4 represent the average strength and modulus vs. fluence for a particular group.

Flexural test results demonstrate a small fluctuation in the mechanical strength and modulus of the composite with various irradiation durations. It is thus believed that the test results revealed no significant changes occurred in the mechanical properties under the experimental conditions used.

Table III. Neutron irradiation effects on the flexural properties of LTM45 composites

Fluence (neutron/cm^2)	0 degree flexure		90 degree flexure	
	Yield stress (MPa)	Modulus (GPa)	Yield stress (MPa)	Modulus (GPa)
Baseline (No radiation)	1194	83.68	68.297	7.48
$1.5x10^{10}$	1166	83.94	64.26	6.92
$3.36x10^{10}$	1232	87.77	66.47	6.97
$6.04x10^{10}$	1179	81.62	64.76	7.32
$1.20x10^{11}$	1150	81.08	72.09	7.45

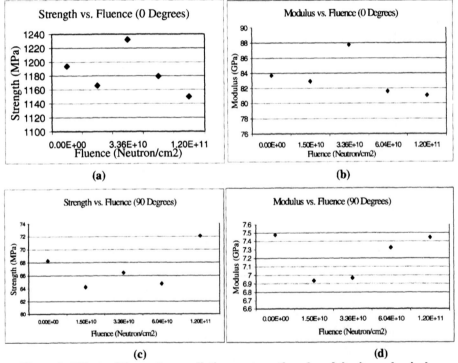

(a)   (b)

(c)   (d)

Figure 4. Effects of the neutron radiation on strength and modulus in mechanical flexural tests of LTM45 composites: (a) & (b), 0°; (c) & (d), 90°.

3.4 Glass Transition Temperature

Glass transition temperature, T_g, is an indication of the molecular structural changes in materials. DSC was used to measure the glass transition temperature of the regolith materials, both before and after radiation exposures. The DSC results are shown in Table IV.

Table IV. T_g of polyimide bonded regolith before and after radiation

Radiation conditions	Glass transition temperature (°C)	
	Regolith 101 Polyimide/regolith 40/60	Regolith 102 Polyimide/regolith 20/80
Baseline (No radiation exposure)	232.92	233.46
55 MeV proton (Fluence: 1×10^6 protons/cm^2)	232.88	233.77
800 MeV proton (Fluence: 5.34×10^9 proton/cm^2)	233.25	234.62
800 MeV proton (Fluence: 1.78×10^{10} proton/cm^2)	231.71	234.22
Neutron (Fluence: 5.0×10^9 neutron/cm^2)	233.29	234.21

It shows that there are no changes in glass transition temperature for both of regolith materials after the radiation exposures under the conditions used in this research work. It is thus believed that there was no large quantity of molecule scission or further crosslinking induced by the radiation in the polyimide. Therefore the thermal properties of the regolith materials would not change after being irradiated. Nevertheless, more direct analysis methods and measurements are needed to determine the radiation effects on the molecular structures of polymers over the longer term irradiation exposures.

For the LTM45 composites, as observed in Figure 5 (a) and (b), DMA results did not show any significant difference between the baseline composite and the irradiated ones. The T_g observed for the baseline panel was 217.83°C, while upon irradiated by 40 MeV protons with a fluence of 1.5×10^{11} protons/cm^2, it was 218.23°C.

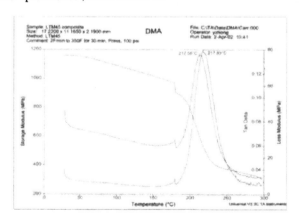

Figure 5. (a) DMA results of LTM45 non-irradiated specimen

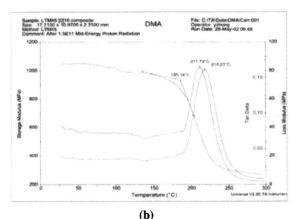

(b)

Figure 5. (b) DMA results of LTM45 specimen irradiated with 40 MeV protons with a fluence of 1.5×10^{11} protons/cm^2.

4. CONCLUSIONS

Polyimide bonded *in-situ* composites with regolith contents of 60% and 80% could shield off and, therefore, reduce the fluence of medium energy proton radiation. But the regolith composites with a thickness of about 2 g/cm^2 did not provide adequate shielding for electronic chips under neutron exposures. The compression strength of the regolith materials changed slightly after neutron and proton exposures. The modulus of the Regolith 101 showed a 16% drop after being irradiated by 800 MeV proton with a fluence of 5.34×10^9 proton/cm^2 and a 20% drop with fluence of 1.78×10^{10} proton/cm^2. The radiation exposures did not have any effect on glass transition temperatures of the regolith materials. Some direct molecular analysis methods should be used to further determine radiation effects on the molecular structures of polymers.

The LTM45 composite materials were used to characterize its space radiation shielding effectiveness and variation in mechanical properties due to the neutron and proton irradiation exposures. A decreasing trend in strength of 0-degree specimens with increased proton irradiation duration and a slight increase in secant modulus of 90-degree specimens with increasing proton fluence were observed. The variation in modulus due to mid-energy proton irradiation was smaller than the changes in strength. Glass transition temperature did not show any significant changes after proton radiation of the materials.

Acknowledgements: This work was funded by NASA through Grant # NCC9-114.

5. REFERENCES

1. "Shielding Strategies for Human Exploration", edited by J. W. Wilson, J. Miller, A. Konradi, and F. A. Cucinotta, NASA Conference Publication #3360, December 1997.
2. J. L. Barth, C. S. Dyer and E. G. Stassinopoulos, "Space, Atmospheric and Terrestrial Radiation Environments", IEEE Trans. Nucl. Sci., Vol. 50 (2003) 466.

3. J. Zhou, J. Moore, H. Huff, S.Ardalan, R. Wilkins, T. Fogarty and Yang Zhong, "Study on regolith material for structural and space radiation shielding applications", 34th International SAMPE Technical Conference, Baltimore, Maryland, Nov. 4 -7, 2002 pp. 358 – 364.
4. R. K. Tripathi, J. W. Wilson, F. A. Cucinotta, J. E. Nealy, M. S. Clowdsley, M.-H. Y. Kim, 01ICES-2326, 2001.
5. M-H Y. Kim, et al, High Perform. Polym., 12 (1), 13 (2000).
6. M-H Y. Kim, et al, SAMPE International Symposium, 44, 359 (1999).
7. V.Calvin, MS Thesis, Prairie View A&M University, 2003.
8. J. Moore, MS Thesis, Prairie View A&M University, 2001.
9. G. Badhwar, H. Huff and R. Wilkins, "Alterations in Dose and Lineal Energy Spectra under Different Shieldings in the Los Alamos High-Energy Neutron Field", Rad. Res. 154, 697-704 (2000).
10. M-H Y. Kim, et al, SAMPE International Symposium, 44, 359 (1999).
11. T. N. Fogarty et al, AIAA-98-0296 (1998).
12. C. L. Snead, Jr., J. Morena, C. J. Czajkowski and J. Skaritka, Materials Characterization 42, 73–91 (1999).

Mater. Res. Soc. Symp. Proc. Vol. 929 © 2006 Materials Research Society   0929-II04-06

Study of the Effects of Various Nanopowders in the Properties of GPC

Renato Amaral Minamisawa, Bopha Chhay, Iulia Muntele, Lawrence Holland, Robert Lee Zimmerman, Claudiu Muntele, and Ila Daryush
Department of Physics, Alabama A&M University, Center of Irradiation of Materials, Post Office Box 1447, Normal, Alabama, 35762

ABSTRACT

We have introduced various nanopowders in the precursor of glassy polymeric carbon (GPC) and studied its electrical, thermal, and mechanical properties as well as its chemical structure. In general the GPC ware produced at AAMU is used for making crucibles, heat exchangers, and for prosthetic devices because of its biocompatibility. GPC ware at AAMU is synthesized from a phenolic resin solution from Georgia Pacific in a pyrolyser system at temperatures between 100 °C all the way to 2800 °C. The heat treatment includes several stages: gelling, curing, postcuring, precarbonization and carbonization. The fabrication of GPC is complicated because of the high production rate of gaseous products in critical temperature ranges where out-diffusion is relatively slow. Special care should be taken in temperature programming to avoid kilning faults and misshapen or porous GPC end results. In this work we have introduced SiC, CNT or Al_2O_3 to the precursor and studied the properties of the final product treated at $1000^\circ C$ pyrolysis temperature.

INTRODUCTION

Glassy polymeric carbon is a typical hard carbon obtained by slow thermal degradation of many thermoset polymers. Its production consists in the slow heat treatment of some hydrocarbon precursor. Above $1000^\circ C$ the material transforms without disruption and with no change in shape into a glass-like polymeric form of almost pure carbon [1]. The long aromatic carbon chains rearranged in a network of microfibrils form a tighter mass, giving good electrical conductivity and mechanical resistance to the material, attributing for the GPC a high performance in extreme environments. Moreover, the strength with exceptional hardness, resistance to thermal shock and several other properties, make GPC unique and much researched for numerous applications in biocompatible prostheses, heart valves, electrodes, crucibles, space research and so forth [2].

More recently, a vast array of applications are emerging that require strategies for nanoparticles dispersion in polymers to improve their properties, such as in the use of fullerenes to enhance the efficiency of polymer photovoltaic devices [3] and mechanical reinforcement of phenolic composites using multi-walled carbon nanotubes [4].

We have conducted a primary investigation to produce and characterize GPC composites doped with silicon carbide (SiC), carbon nanotube (CNT) and alumina (Al_2O_3) nanopowders in order to transfer some of the powder properties to the GPC. Carbon nanotubes with excellent electrical conductivity, and mechanical and chemical resistance have been reported in several applications [5]. Alumina is known as a good thermal and electrical insulator and SiC as a semiconductor with high thermal conductivity. The method developed to disperse the nanoparticles in the phenolic resin precursor and the electrical, thermal and mechanical properties such as the chemical characterization are reported.

EXPERIMENTAL PROCEDURES

We chose multi-walled carbon nanotubes (CNTs), 50 nm alumina and silicon carbide (SiC) nanopowders as doping materials for GPC due to their specific electrical, thermal and mechanical properties. The GPC precursor used was phenolic resin from Georgia Pacific.

Nanopowders were placed separately in containers containing ethanol to separate the aggregation that could cause nonhomogeneous dispersion in the GPC precursor. Phenolic resin was poured into the containers, infiltrating the nanoparticles. The solution was mixed slowly and sonicated, to eliminate air bubbles generated in the process, and then poured in onto a glass mold frame with volume of 55 x 20 x 1 mm^3. Diluting the solution with ethanol reduces the occurrence of trapped bubbles during resin polymerization, which occurs mainly at the surface. The added nanoparticles may offer surfaces for nucleation of bubbles. Dilution with ethanol delays the polymerization process, possibly reducing bubble entrapment.

To speed evaporation, the samples were treated at a pressure of about two thirds of an atmosphere for one day. The samples with 1.0 wt% nanopowder were then gelled in an oven at 60°C for four days. They were then separated from the mold, to be finally pyrolyzed up to 1000°C in argon. Mechanical tests of the pyrolized GPC:nanopowder composites were performed using a strain-stress testing machine and electrical resistivity was measured by the Van der Pauw method. The thermal stability of GPC:nanopowder composites was tested exposing the samples at 1000°C in forming gas, 4% H$_2$ in argon. We investigated the chemical and structural characteristics of the composites using Rutherford Backscattering Spectroscopy (RBS), Raman and FTIR spectroscopy.

RESULTS AND DISCUSSION

Chemical structure analysis

Figure 1 shows the attenuated total reflectance (ATR) FTIR spectra of GPC:nanopowder composites.

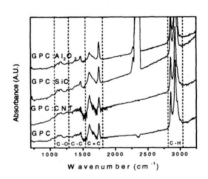

Figure 1. ATR analysis of GPC:nanopowder composites.

The expected C=C and C-C bond peaks of GPC chains are found in all the spectra. However, the C-O and C-H bonds are still present, indicating that 1000°C heat treatment temperature (HTT) is not enough to eliminate all the oxygen and hydrogen atoms from the GPC. Also, as ATR analysis measures the infrared absorption at the sample surface, some C-O and C-H bonds could be formed in air after pyrolization. No signals from SiC or alumina nanoparticles are observed, which suggests that these bonds are not activated by the infrared stimulation.

The RBS analysis shown in figure 2 presents the atomic composition of each GPC:nanopowder composite. A higher concentration of oxygen atoms was observed in the surface, verifying the ATR measurements. Some contamination, such as Ca and Si atoms, were found in all the samples, accumulated during sample preparation and pyrolisis.

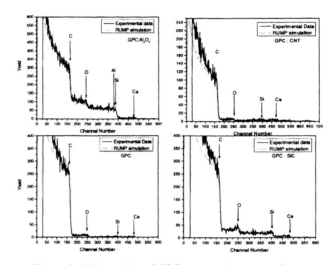

Figure 2. RBS analysis of GPC:nanopowder composites.

The D and G lines in the Raman spectrum were verified for all the samples (figure 3) and the D over G ratio calculated in order to determine the structural state of the material. As in other sp_2 carbon materials, the first order Raman spectra of pyrocarbons exhibit the degree of amorphous structural disorder in the D line, and the degree of graphitization in the G line. The ratio of the areas of the lorentzian fits in each peak was used to calculate the D/G ratios for GPC, GPC:SiC, GPC:alumina and GPC:CNT, obtaining respectively 2.1, 1.98, 2.02 and 2.03. From the results we infer that the nanopowders promote organization of the graphitization at the material surface, acting as catalysts during the pyrolization process.

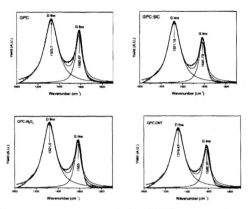

Figure 3. Raman spectroscopy analysis of GPC:nanopowder composites 1.0 wt%.

Mechanical tests

Figure 4 shows the typical experimental strain-stress curve of the GPC:nanopowder composites, the Young's modulus and the fracture strains are presented in table I.

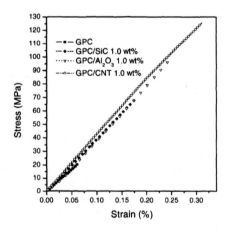

Figure 4. Mechanical tests o f GPC:nanopowder composites HTT at 1000°C.

Composite	Young's Modulus	Maximum Force	Strain
	(GPa)	(MPa)	%
GPC	33.53	24.44	0.06
GPC:SiC	38.46	72.34	0.18
GPC:alumina	40.96	102.76	0.25
GPC:CNT	41.44	128.25	0.31

Table I. Experimental mechanical properties of GPC:nanopowder composites.

Significant variations in failure strain were observed for all the GPC:nanopowder composites when compared with the GPC sample without nanoparticles. The GPC:CNT composite shows the higher mechanical resistance when compared with the other compounds. However, is important to highlight that this behavior was observed for samples prepared with 1.0 wt% of nanopowders, and can change for different concentrations.

Thermal exposure test

We made Raman spectroscopy analyses of the GPC:nanopowder composites, annealed for 2 hours at 1000°C in forming gas (argon:4%H_2), in order to verify the chemical stability. The D over G ratio for all samples before and after annealing is shown on table III. The calculations are based on the ratios of the lorentzian fit D peak area over that of the G peak area. We also checked for mass losses.

Table III. Raman analysis of GPC samples exposed to high temperature in H_2 forming gas.

Sample	Mass before heating (g)	Mass after heating (g)	D/G before exposure	D/G after exposure	D/G variation Δ(%)
GPC	0.28 ±0.01	0.27 ±0.01	2.02	1.97	-2.5
GPC:SiC	0.53 ±0.01	0.53 ±0.01	1.93	1.99	+3.0
GPC:CNT	0.58 ±0.01	0.58 ±0.01	1.96	1.96	0
GPC:alumina	0.43 ±0.01	0.42 ±0.01	1.96	1.97	+0.5

The GPC samples show negligible variation in D/G, which suggests high thermal stability in the material. GPC without doping may show slight graphitization while the samples with alumina and SiC suggest an unexpected slight increase in the structural disorder of the material. There is no measurable mass loss.

Electrical measurements

The average values and the standard deviations for electrical resistivity ρ of different GPC:1wt% nanopowder composites, with HTT 1000°C, are shown in table II. The uncertainties and errors associated in these measurements have many components related, mainly with the measurements of the sample resistance and dimensions. The close values of ρ and the high standard deviation suggests that GPC conductivity for 1000°C is relatively good and that the nanoparticle concentration is not enough to influence the current passage in the microfibril network of the carbon.

Table II. Electrical resistivity measurements of GPC:nanopowder composites.

Compound	Resistivity ρ $(x10^{-3} \text{ ohm-cm})$	Standard Deviation $(x10^{-3} \text{ ohm-cm})$
GPC	5.09	0.28
GPC:SiC	4.89	0.1
GPC:CNT	4.84	0.22
GPC:alumina	5.01	0.25

CONCLUSION

We have produced and characterized composites of GPC doped with different nanopowders. We have developed a method for sample preparation without voids using dilute solutions of phenolic resin in ethanol to mold precursors with nanopowders. We report preliminary tests of electrical, thermal and mechanical properties, as well as chemical analysis. The GPC samples were doped 1wt% with alumina, CNT or SiC and fired to 1000°C HTT. In conclusion, nanoparticles at this level added to the GPC precursor affect only the mechanical properties of GPC, showing a potential for utility. Further investigation might be explored in the future for a better understanding of the influence of nanopowders on GPC pyrolisis and their effects on the polymer properties.

ACKNOWLEDGMENTS

This research is sponsored by the Center for Irradiation of Materials, Alabama A&M University and by the AAMURI Center for Advanced Propulsion Materials under the contract number NAG8-1933 from NASA, and by National Science Foundation under Grant No. EPS-0447675.

REFERENCES

1. R. L. Zimmerman, D. Ila, D. B. Poker, S. P. Withrow, *Nucl. Instr. Meth.* **B127/128** 1023 (1997).

2. G. M. Jenkins, K. Kawamura, (Polymeric carbons – carbon fibre, glass and char, Cambridge University Press, 1976).

3. N. S. Sariciftci, L. Smilowitz, A. J. Heerger, F. Wudl, *Science* **258**, 1474 (1992).

4. M. K. Yeh, N. H. Tai, J. H. Liu, *Carbon* **44**, 1 (2006).

5. M. F. Yu, O. Lourie, K. Moloni, T. F. Kelly, R.S. Ruoff, *Science* **287**, 637 (2000).

Mater. Res. Soc. Symp. Proc. Vol. 929 © 2006 Materials Research Society   0929-II04-08

Chemical, Mechanical and Electrical Properties of Glassy Polymeric Carbon

Iulia C. Muntele, Claudiu I. Muntele, Renato Minamisawa, Bopha Chhay, and Daryush Ila
Center for Irradiation of Materials, Alabama A&M University, 4900 Meridian Street, Normal, AL, 35762

ABSTRACT

Glassy Polymeric Carbon (GPC) is obtained by a molding technique, in various shapes, from a phenolic resin precursor. The heat treatment of the precursor is achieved in three stages up to 1000 °C. Similar GPC materials produced in our laboratory displayed large strain to failure ratio, small thermal expansion coefficient and low density. Like all carbon forms, is attacked by oxygen, especially atomic oxygen. Nevertheless the kinetics for reaction with atmospheric oxygen is very slow. We investigated the composition and structural changes of the phenolic precursor as a function of temperature and evaluated material's stability when exposed to high temperatures in presence of hydrogen (H_2) or oxygen (O_2).

INTRODUCTION

GPC is a polymer obtained by pyrolysis of resols (in our case formaldehyde resin provided by Georgia Pacific, GP 387G84 Resi-Lam Phenolic Laminating Resin). GPC is a unique material, relatively chemically inert (at room temperature), and biocompatible [1]. GPC can withstand high temperature and corrosive environments, with applications in nuclear reactors and space industry (nuclear fuel cells, heat shied for thermal protection systems). A vast amount of literature [1-5] is available in regard with the use of heat-treated phenolic resin and its applications in construction industry for coatings, adhesives, polymer flame-retardants, and composites. Some desirable properties of this material are low density (1 g/cm^3), low thermal conductivity (0.238-1.428 W/mK), low thermal expansion coefficient (10^{-8} K^{-1}), high critical oxygen index (difficult to ignite and maintain a burning condition) [6]. One indication of material's thermal stability can be inferred from the fact that its chemical kinetics at any given temperature moves to completion very quickly and subsequent exposure at the same temperature will not induce further structural change [1]. This thermal stability suggests that the material will retain its structural integrity and mechanical and physical properties under adverse conditions. The objective of this study was to obtain and characterize the composition and structural changes of pure GPC samples in extreme environments.

EXPERIMENT

GPC sample preparation: liquid resol was diluted 1:1 dilution ratio with ethyl alcohol, sonicated, kept 3 hours at gelling temperature, molded in the desired shape and gelled for 5 days at 75 °C. After removal from mold the samples are pyrolyzed up to 1000 °C according with a designed heat treatment [7] program for 170 hours. Samples prepared at different heat treatment stages (75, 200, 500, 700 and 1000 °C) were characterized using Rutherford Backscattering Spectrometry (RBS), FTIR/ATR Spectroscopy, Raman Spectroscopy, Residual Gas Analysis (RGA), mechanical characterization, and resistivity measurements. RBS analysis was done with 2.1 MeV alpha particles from a Pelletron accelerator, detected at 170° using a SSB detector with 20 keV FWHM. FTIR/ATR was done with a Tensor 27 spectrophotometer with a Pike-ATR accessory equipped with a ZnSe crystal. Raman spectra were acquired with a LabRam spectrophotometer using a He-Ne laser with λ=632 nm. RGA analysis was done in a vacuum chamber with a SRS 200 residual gas analyzer. The stress-strain curve was obtained on a Zwick/Roell instrument. Resistivity measurements were done using the Van de Paaw method on an MMR Technology setup.

RESULTS AND DISCUSSION

Figure 1 shows the RBS spectra of the resol exposed successively at room temperature, 75 °C, 200 °C, 700 °C and 1000 °C. The main constituents of the samples are C and O. However, surface contamination in form of Si, Ca and O is present, the source of the contamination being traced to the soda-lime microscope slide used in the mold. The amount of contaminats is decreasing as the heat treatment reaches completion at 1000 °C. The oxygen content in the bulk of the sample is decreasing with the increase in temperature from approximately 10 at% when gelled to less than 1 at% after pyrolysis is completed at 1000 °C.

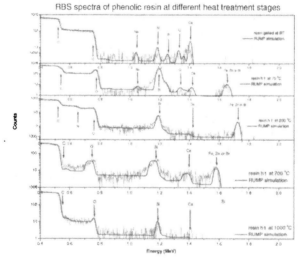

Figure 1: RBS spectra of various heat treatment stages. Oxygen content in the bulk is decreasing with the heat treatment from 10 at% at 75 °C to <1 at% at 1000 °C.

Figure 2 (left) is a succession of FTIR-ATR spectra showing the structural change of the resol with the increase in temperature: as temperature evolves the methylene bridges rupture, the C-O-C infrared band is reduced in intensity, and oxygen is removed from the polymeric chains. As temperature increases past 400 °C, hydrogen is also released. A detailed account of the polymerization process is given by [1]. Figure 2 (right) shows the increase in the G and D bands at different heat treatment stages. Between gelling temperature (75 °C) and 200 °C there is a broad band around 1100 cm^{-1}. The Raman spectrum for the 200 °C treatment shows an incipient G band. At 500 °C the D and G bands are well defined. The broad band at 1100 cm^{-1} is still very strong; this band is attributed to amorphous (hydrogenated) carbon [8-10]. After 700 °C heat treatment the 1100 cm^{-1} band is shifted to the right and is highly decreased, which corresponds with the hydrogen removal from the sample as described in literature [1, 7].

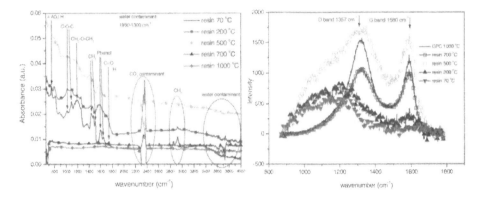

Figure 2. (left) FTIR-ATR spectra of resol samples at different heat treatment stages; (right) Raman spectra of resol sample at different heat treatment stages.

Figure 3 (left) shows the evolution of the GPC sample (pyrolyzed at 1000 °C) after exposure to forming gas (Ar+4%H$_2$) and residual air at 400, 500 and 600 °C in one-hour intervals. Figure 3 (right) shows the evolution of the GPC sample (pyrolyzed at 1000 °C) after exposure only to forming gas at 1000 °C for 30 minutes and 2 hours respectively. We choose to monitor the Raman signal from the G and D lines as the treatment was applied. In the case of the samples exposed to residual molecular oxygen we noticed a decrease in the D:G ratio. Moreover, by eye inspection we could observe changes on the surface of the sample. The strained surface layer cracked and the residual oxygen had access to free carbon bonds, hence burning the sample. The 1cm^2 area sample burned completely after exposure to molecular oxygen for 1 hour at 800 °C. In the case of the sample exposed only to forming gas, there is no consistent change in the D:G ratio. There is a decrease in the intensity of the two lines, which can be explained by reorientation of the C bonds due to stress release at elevated temperature. There was no change in the mass of the sample before and after exposure to forming gas.

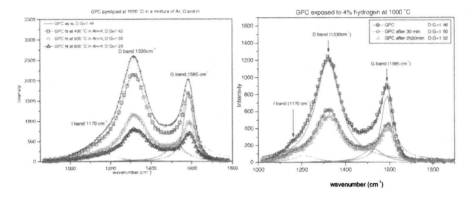

Figure 3: Figure 3 (left) shows the evolution of the GPC sample (pyrolyzed at 1000 °C) after exposure to forming gas and residual air at 400, 500 and 600 °C in one-hour intervals. Figure 3 (right) shows the evolution of the GPC sample (pyrolyzed at 1000 °C) after exposure to forming gas only for 30 minutes and 2 hours respectively at 1000 °C.

Figure 4 shows the stress-strain curves for gelled and pyrolyzed resin. From the figure and table 1 one can see that by pyrolysis the resin has a higher Young's modulus, can take a higher load than the gelled resin, and is more brittle.

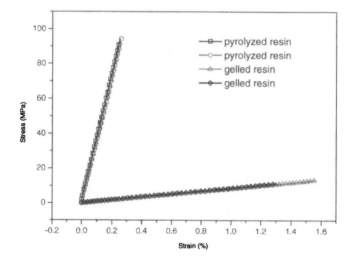

Figure 4: Stress-stain curves for gelled and resin pyrolyzed at 1000 °C.

Table 1: Flexural test results for gelled and pyrolyzed resin.

	Sample ID	E-Modulus	Fmax.	ε Break
Nr		GPa	MPa	%
1	Gelled resin	0.87	13.30	1.57
2	Gelled resin	0.85	11.42	1.30
3	Pyrolyzed resin	36.73	94.20	0.25
5	Pyrolyzed resin	35.67	96.07	0.27

RGA measurements in vacuum at 1000 °C don't show any emission above background from the pyrolyzed sample. This is due to the fact that at any given temperature the kinetics of chemical reactions is very fast [1], hence subsequent exposure to the same temperature will not induce further polymerization, and this is way the RGA measurements done in vacuum at 1000 °C will not show additional gas removal from the sample, or mass change.

For resistivity measurements, flat resin samples 1 mm thick and 1 cm^2 area, heat treated at different temperatures, were used for resistivity measurements. Precautions were taken to follow the guidelines for the Van de Pauw method. The contacts are on the boundary of the sample and the contact surface is very small. The sample is thin relative to the other dimensions and there are no isolated holes within the sample. The measured resistivity values are:

Gelled resin: $\rho = 5.4 \times 10^3$ Ωm – this value is highly underestimated.

Resin post-carbonized at 500 oC: $\rho = 4.8 \times 10^3$ Ωm

GPC: $\rho = 0.3$ Ωm

The resistivity values for the resin heat treated above gelling temperature correspond as order of magnitude with the data presented in the related literature.

CONCLUSIONS

Using a molding technique combined with pyrolysis, we obtained glassy polymeric carbon. Using FTIR and Raman we monitored the chemical and structural changes of this material at different heat treatment stages. We have shown that pure GPC pyrolyzed at 1000 °C burns in air at a slow rate (1 cm × 1 cm × 1mm sample burns completely in 1 hour at 800 °C). As expected, Raman analysis shows an increase in the amorphous carbon content with increase in the temperature in the presence of molecular oxygen. Heat treatment in vacuum up to the temperature of pyrolysis shows no degradation of the sample. Preliminary mechanical tests (engineering stress-strain curve) show promising mechanical properties. Electrical resistivity is high in gelled and post-carbonized resin samples, and low in GPC samples pyrolyzed at 1000 °C. RBS analysis of the trace impurities show surface contamination with silicon and calcium for pyrolyzed samples. The source of this contamination can be traced to the soda-lime glass used as mold. Exposure to molecular hydrogen at 1000 °C is not inducing changes in the chemical structure of the sample and the sample's mass remains unchanged.

Overall, glassy polymeric carbon (GPC) is a lightweight material, which seems to be suitable for use in corrosive atmosphere at elevated temperatures for limited amounts of time.

ACKNOWLEDGMENTS

Research sponsored by the Center for Irradiation of Materials, Alabama A&M University and by the AAMURI Center for Advanced Propulsion Materials under the contract number NAG8-1933 from NASA, and by National Science Foundation under Grant No. EPS-0447675.

REFERENCES

1. G. M. Jenkind, K. Kawamura, in Polymeric carbons-carbon fibre, glass and char, Cambridge University Press, Cambridge, 1976, p. 68.
2. H.-T. Chiu, S.-H. Chiu, R.-E. Jeng, J.-S. Chung, Polymer Degradation and Stability 70 (2000) 505-514.
3. J. Bijwe, Nidhi, N. Majumdar, B.K. Satapathy, Wear 259 (2005) 1068–1078
4. Chin-Lung Chiang, Chen-Chi M. Ma, Polymer Degradation and Stability 83 (2004) 207–214.
5. C.P. Reghunadhan Nair, R.L. Bindu, K.N. Ninan, Polymer Degradation and Stability 73 (2001) 251–257.
6. S. G. Kuzak, J. A. Hiltz, P. A. Waitkus, Journal of Applied Polymer Science, Volume 67, Issue 2, Pages 349 – 361.
7. H. Maleki, L. R. Holland, G. M. Jenkins and R. L. Zimmerman, Carbon Vol. 35, No. 2, 227-234, 1997.
8. M.A. Tamor, W. C. Vassell, J. of App. Phy, vol. 76, no. 6, p. 3823-3830, 1994.
9. J. Schwan, S. Ulrich, V. Batori, H. Ehrhardt, J. of Appl. Phy., vol. 80, issue 1, pp. 440-447.
10. J.-M. Vallerot, X. Bourrat, A. Mouchon, G. Chollon, Carbon 44 (2006) 1833-1844.

Environmental Effects
(Temperature, Radiation, Corrosion, Erosion, and Pressure)

Mater. Res. Soc. Symp. Proc. Vol. 929 © 2006 Materials Research Society 0929-II05-02

Metallurgical and Corrosion Studies of Modified T91 Grade Steel

Pankaj Kumar[1], Debajyoti Maitra[1], and Ajit K Roy[2]

[1]Mechanical Engineering, University of Nevada Las Vegas, 1173 Maryland Circle Apt. #3, Las Vegas, NV, 89119

[2]Mechanical Engineering Department, 4505 Maryland Parkway, University of Nevada Las Vegas, Las Vegas, NV, 89154

ABSTRACT

Modified 9Cr-1Mo steels containing different Silicon (Si) content have been tested for evaluation of their metallurgical and corrosion properties. The results of tensile testing indicated reduced failure strain up to 400°C followed by its enhancement at a higher temperature. Stress corrosion cracking tests at constant load did not show any failure in any materials. However, reduced ductility and true failure stress were noted in slow strain rate (SSR) testing, indicating increased cracking susceptibility at elevated temperatures. The corrosion potential became more active with increasing temperature in polarization studies. A combination of intergranular and cleavage failures was observed in specimens tested by the SSR technique.

INTRODUCTION

Transmutation of spent nuclear fuel is receiving significant attention during this past decade to circumvent the problems associated with their disposal. Transmutation refers to the transformation of isotopes with high radioactivity and long half-lives into species with reduced radioactivity and shorter half-lives through separation and decay of minor actinides and fission products. During this process the spallation target material will be contained in a vessel made of suitable structural material such as modified 9Cr-1Mo steel, also known as T91 grade steel. The target material is currently being envisioned is a molten lead-bismuth eutectic having a temperature ranging between 425 and 550°C. Since the T91 grade steel will be subjected to this temperature regime during spallation process, efforts have been made in this investigation to determine the tensile properties of modified 9Cr-1Mo steel containing Si ranging from 0.5 to 2 weight%. The addition of Si in this type of martensitic grade steel has been proved to be beneficial in a previous study performed by the authors.

The susceptibility of these materials to stress corrosion cracking (SCC) has also been evaluated in an aqueous solution to develop a baseline SCC data as a function of Si content at different temperatures. SCC behavior in a molten LBE is also currently under investigation. Simultaneously, the localized corrosion behavior of these alloys has been studied by an electrochemical technique in an identical aqueous solution. The morphology of failure of the cylindrical specimens used in SCC testing has also been determined by Scanning Electron Microscopy (SEM). This paper presents the overall results showing the effect of Si content on both the metallurgical and corrosion behavior of modified T91 grade steel.

EXPERIMENTAL PROCEDURE

Modified T91 grade steels containing 0.5, 1.0, 1.5 and 2.0 wt% of Si were custom melted by vacuum induction melting practice. Their compositions are given in Table1. These experimental heats were subsequently processed by forging and hot rolling into the desired shapes. The processed materials were then heat treated to achieve the desired metallurgical microstructures. They were austenitized at 1850°F for 1 hour followed by an oil quench thus producing brittle martensite. Subsequently they were tempered at 1150°F for 1 hour and air cooled. This thermal treatment resulted in fully tempered and fine grained martensitic microstructure, as shown in figure 1.

Table 1: Chemical Compositions of Four Heats of T91 Grade Steel

Material/ Heat No.	Elements (wt %)												
	C	Mn	P	S	Si	Ni	Cr	Mo	Al	V	Cb	N(ppm)	Fe
T91/ 2403	.12	.44	.004	.003	.48	.30	9.38	1.03	.024	.23	.91	570	Bal
T91/ 2404	.12	.45	.004	.003	.02	.30	9.61	1.03	.025	.24	.89	529	Bal
T91/ 2405	.11	.45	.004	.004	.55	.31	9.66	1.02	.024	.24	.085	485	Bal
T91/ 2406	.11	.45	.004	.004	.88	.31	9.57	1.01	.029	.24	.087	302	Bal

Bal: Balance

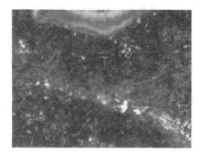

Figure 1: Optical Micrograph of T91 (1%Si) grade steel, 500X

    Smooth cylindrical specimens were machined from the heat treated rectangular bars in such a way that the gage section was parallel to the longitudinal rolling direction. A ratio of 4 was maintained between the gage length and the gage diameter according to the ASTM Designation E8 [1]. The tensile properties involving these cylindrical specimens were determined at ambient temperature, 150, 300, 400 and 550 °C in an Instron testing equipment containing a customized ceramic chamber. For high temperature testing N_2 was flown inside the test chamber to prevent oxidation of the specimens. The tensile properties were determined in the form of engineering stress vs. strain (s-e) diagrams as a function of temperature.

    The susceptibility of all four heats to SCC was determined using cylindrical specimens under both constant load and SSR conditions. A strain rate of $3.3 \times 10^{-6} \, s^{-1}$ was used in SSR testing. The cracking susceptibility at constant load was determined in terms of the threshold stress below which no failure occurred in the aqueous solution. For SSR testing, the ductility parameters including percent elongation (%el) and percent reduction in area (%RA), true failure stress (σ_f) and time to failure (TTF) were used to characterize the cracking tendency.

For localized corrosion studies, cyclic potentiodynamic polarization (CPP) method was used that was based on a three electrode polarization concept. The test specimen was used as a working electrode, two graphite rods acted as cathodes and Ag/AgCl worked as the reference electrode. The potentiostat was calibrated prior to the CPP testing according to ASTM Designation G5 [2]. The extent and morphology of failure of the tested cylindrical specimens were determined by SEM.

RESULTS AND DISCUSSION

The results of tensile testing for T91 grade steel containing 1 and 2 wt% Si are shown in figures 2 and 3, respectively. An evaluation of the s-e diagrams, superimposed for different temperatures, indicate that there was a gradual reduction in failure strain (e_f) for steel containing 2 wt% Si at temperatures up to 400°C. However above 400°C, the failure strain was enhanced as shown in figure 3. As to the s-e diagrams for T91 grade steel containing 1 wt% Si, very little or no variation in e_f was noted due to change in temperature. The phenomenon of reduced e_f in a temperature range as noted in this investigation may possibly be attributed to Dynamic Strain Ageing (DSA) associated with pinning of dislocation by interstitial solute elements in the vicinity of the grain boundaries [3]. The variation in Si content may also play an important role in DSA of martensitic material, which is currently being studied by Transmission Electron Microscopy (TEM). As anticipated, the magnitudes of yield strength (YS), ultimate tensile strength (UTS) and failure stress were gradually reduced with increasing temperature due to enhanced plastic flow.

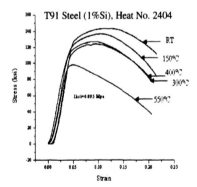

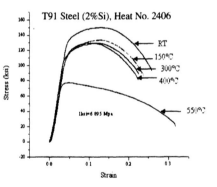

Figure 2: s-e Diagram versus Temperature for T91 grade steel with 1% Si content

Figure 3: s-e Diagram versus Temperature for T91 grade steel with 2% Si content

RT: Room Temperature

RT: Room Temperature

The results of SCC testing indicate that no failure was observed at constant load irrespective of the Si content even at applied stresses corresponding to 95% of the materials yield strength values. Thus a threshold stress equivalent to 0.95YS can be used for these alloys when tested in a similar aqueous environment. The cracking susceptibility under SSR condition is illustrated in figures 4 and 5 in the form of s-e diagram as a function of temperature. The magnitudes of ductility (%el and %RA), σ_f and TTF determined from these s-e diagrams and specimen dimensions before and after testing are given in Table 2. The variations of all four parameters with temperature are illustrated in figure 6, indicating enhanced cracking tendency at elevated temperatures. The results of CPP testing exhibited classical active to passive behavior with a positive hysteresis loop, as shown in figure 7. The corrosion potential became more active at 60°C compared to that at 30°C. A similar effect of temperature on E_{corr} has been sited in the open literature. The fractographic evaluation by SEM revealed dimpled microstructures as shown in figure 8 in the cylindrical specimens tested in tension at different temperatures, indicating predominantly ductile failure. However the cylindrical specimens subjected to SSR testing showed a combination of intergranular and cleavage failure as depicted in figure 9, that are the characteristics of brittle failure.

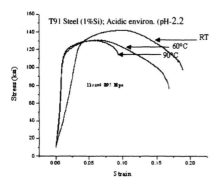

Figure 4: s-e Diagram vs Temperature

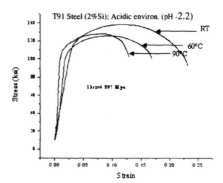

Figure 5: s-e Diagram vs. Temperature

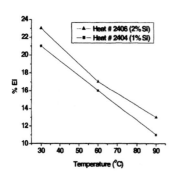

Figure 6 (a): %El vs. Temperature

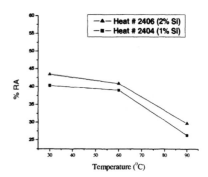

Figure 6 (b): %RA vs. Temperature

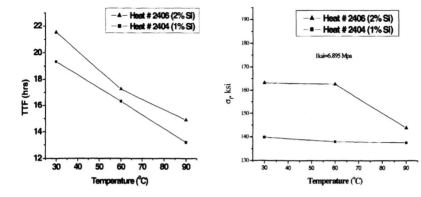

Figure 6(c): TTF vs. Temperature   Figure 6(d): True Failure Stress vs. Temperature

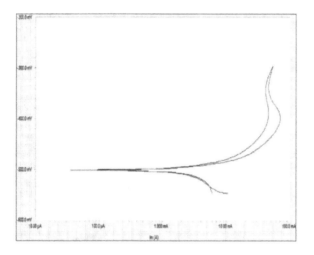

Figure 7: CPP Diagram of T91 Grade Steel
Containing 1 wt % Si

 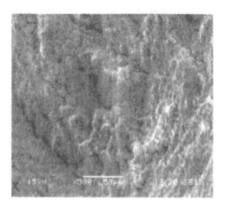

Figure 8: SEM Micrograph of T91 Grade
Steel (2% Si) of at 550°C

Figure 9: SEM Micrograph of T91 Grade
Steel (1% Si) after SSR testing at 60°C

SUMMARY AND CONCLUSIONS

T91 grade steels containing 4 levels of Si content were characterized for metallurgical and corrosion behavior for application as target materials in transmutation of SNF. The key results are summarized below:

- Steel containing higher Si content exhibited reduced failure strain possibly due to DSA behavior.

- No failure was noted in constant load SCC testing. However reduced ductility, σ_f and TTF were observed at elevated temperatures indicating enhanced cracking susceptibility.

- Ductile failure characterized by dimples was seen in tensile testing. However, combined intergranular and cleavage failure were observed in specimens used in SSR testing.

REFERENCES

[1] ASTM Designation E 8, "Standard test methods for tensile testing of metallic materials"
[2] ASTM Designation G 5, "Standard Test Methods for Tensile Testing of Metallic Materials".
[3] R. Kishore, R.N Singh and B.P Kashyap, Journal of Materials Science 32 (1997) pp. 437-442

Mater. Res. Soc. Symp. Proc. Vol. 929 © 2006 Materials Research Society   0929-II05-03

A Novel Method for the Diffusion of Boron in 60-80 Micron Size Natural Diamond Type II/A Powder

Adrian E. Mendez[1], Mark A. Prelas[1], Michael Glascock[2], and Tushar K. Ghosh[1]
[1]Nuclear Science and Engineering Institute, University of Missouri-Columbia, E2433 Engineering Building East, Columbia, MO, 65211
[2]University of Missouri-Columbia, Columbia, MO, 65211

ABSTRACT

The purpose of this paper is to report the experimental results of boron doping on 60-80 micron size diamond particles using Field Enhanced Diffusion with Optical activation (FEDOA) [1-4]. Diamond is a wide band gap material with unique combinations of optical, thermal, mechanical and electronic properties that can be useful for a number of applications including optoelectronic applications and micro sensor technology. The incorporation of boron into diamond has been proven to change its electrical properties and convert the diamond from insulator to a p-type semiconductor [3]. A promising technique for incorporation of impurities into diamond is FEDOA. FEDOA drives impurities into single crystalline diamond material and this method has been demonstrated using a number of impurities [5-7]. The method uses a combination of thermal diffusion with bias, thermal ionization and optical ionization in one setup. A modified version of FEDOA was implemented for the diffusion of Boron in natural Diamond type II/a powder of size 60-80 microns (Figure 1). The diamond powder was obtained from Microdiamant with 99.9% purity. The boron powder used in the experiment was amorphous, 325 mesh 90%(Assay), Mg (5%) nominal obtained from AESAR. A mixture of 3:1 boron powder to diamond powder was used. A heating element and a powder holder were designed and incorporated in the FEDOA system. Electron Microscopy (SEM), Energy-Dispersive X-ray Spectroscopy (EDS) are used to study the diamond-doped morphology and identify impurities. Boron and hydrogen concentration results in the doped samples were studied using Prompt Gamma Neutron Activation Analysis (PGNAA) at the University of Missouri Research Reactor (MURR). Raman analysis of the treated samples is also shown as part of this work. The experimental results show that the samples were effectively doped with boron. It was also found that samples with high boron concentration exhibit high conductivity. This work presents strong evidence that boron can be diffused into natural diamond powders. It also demonstrates that the FEDOA diffusion process is not only a powerful technique for the diffusion of impurities into wide band-gap materials in the form of single crystal plates and polycrystalline plate but also in a powder form with the modified FEDOA process.

INTRODUCTION

The development of new technology based on diamond provides a promising future due to its ability to be adapted to several conditions such as radioactive environments and space exploration. Diamond is a wide band-gap material with unique combinations of optical, thermal, mechanical and electronic properties that can be useful for a number of applications: in harsh environments, under extreme temperatures, during exposure to intense radiation, and in chemically corrosive environments. These exceptional properties suggest the potential of diamond as a material useful in severe environments where present state-of-the-art semiconductor devices are not practical or effective. The recently proven ability to diffuse several elements in natural diamond that can act as donors or acceptors [5-7] has opened the possibility of developing micro-electronic applications such as micro-sensors and LED's that can tolerate such extreme environments. Diffusion is the movement of particles due to concentration gradients from a zone of higher concentration to a zone with lower concentration, as the result of thermal motion of atoms. FEDOA is a powerful technique that enhances diffusion of impurities into wide band gaps materials developed by our group at the University of Missouri-Columbia. Diffusion of impurities into single crystalline diamond material has been studied using this method [3-6], which uses a combination of heat, electric fields, thermal ionization and optical ionization in one setup. The influence of an electric field on diffusion helps to drift selectively ionic species through the diamond. Optical ionization is used to help to ionize impurities that cannot be ionized thermally and to lengthen ion lifetimes. Laser energy absorbed directly by the diamond lattice will also be beneficial for diffusion, as the lattice will lose some of its rigidity. The introduction of the optical field along with heat and the electric field was able to enhance diffusion by two to four orders of magnitude depending upon the impurity [1-4].

Boron is a relatively shallow acceptor in the diamond band-gap (ionization energy of boron level $\Delta E = 0.37$ ev)[7]. It is known that the diffusion of boron in diamond can change its electrical and optical properties. It changes diamond from being an insulator having a high resistivity (10^{13} - 10^{16} Ohm/cm) to being a semiconductor with p- type characteristic[7]. It also changes the transparency of diamond and produces blue to dark color on diamond depending on the concentration of boron. Previous experiment using FEDOA on natural diamond plates have been shown to increase the diffusion of boron in diamond [1-3]. In this study boron is incorporated in diamond powder using the FEDOA concept.

EXPERIMENTAL DETAILS

In order to diffuse boron in diamond particles a modified FEDOA method was developed. In previous work [4-6], using a modification of FEDOA, it was found that certain elements diffused as acceptors or donors. In this work several elements have been found to be able to diffuse in diamond plates. A key to the method is that elements used had ion sizes on the order of the diamond lattice spacing of 1.54 Angstroms. Other criteria were that compounds containing the elements which are used for the FEDOA process have melting points in the range of ~500 °C to 1200 °C . In the present work we studied the diffusion process in diamond particles of 60-80 micron size. The modified version of FEDOA was implemented for the study of diffusion of boron in diamond powder. The modified system is shown in Figure 1.

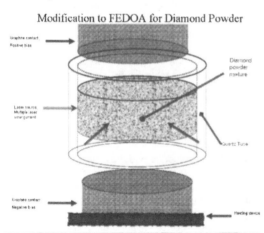

Modification to FEDOA for Diamond Powder

An IR and EDS analysis was performed on the diamond powder before treatment to identify impurities in the diamond. No impurities were found in the diamond at the detection sensitivity level of the instrument.

A. Mixture preparation - Amorphous boron powder was selected as the impurity for diffusion.

Figure 1. Setup for Field Enhanced Diffusion with Optical Activation (FEDOA) Experiments

The boron powder used in the experiment was amorphous, 325 mesh 90% (Assay), Mg (5%) nominal obtained from AESAR. EDS analysis was performed on the boron powder to identify additional impurities. A mixture of 3:1 boron powder to diamond powder was used. The powders were mixed in a mortar and pestle to form a homogeneous mixture. Approximately 0.8g of diamond was used for each run. The mixture was lightly packed inside a quartz vial. Two graphite electrodes made the electrical contact at each end of the quartz tube. One graphite-electrical contact has a positive bias and the other one had a negative bias with a potential difference of 150 V. A laser illuminated the sample. The treatment was performed in the experimental chamber using hydrogen atmosphere under a pressure in the range of 20 to 40 mmHg. Treatment temperature and time are described in the experimental matrix. After the treatment, the sample was allowed to cool to room temperature inside the experimental chamber. All the samples used the same boron-diamond mixture.

B. Cleaning Procedure - A thorough cleaning procedure was performed on the treated sample to remove any residue or "dust" left from the treatment procedure on the surfaces. The treated diamond powder was cleaned using the following steps: (1)The sample was placed in a solution of 200 ml of de-ionized water and 100 mL of HNO_3 to dissolve boron for 30 minutes then the solution was removed by pipetting;(2) a 1:1 mixture of de-ionized water with HNO_3 was added and heated to a boil for 1 hr. (3) the sample was further cleaned in a diluted boiling solution of sulfuric acid (96.6% assay) and chromic acid for 60 min to remove any other impurities that might present on the surface; (4) the sample was rinsed in Aqua-Regia and then in de-ionized water and filtered using a micron size membrane and filtration apparatus; (5) finally the sample was dried in an oven at $100\ C^{o}$ for 30 minutes

C. Impurity and Surface analysis - Each cleaned sample was observed under a transmitted light microscope at magnifications from 20X to 600X. SEM tomography and impurities analysis was performed using an AMRAY 1600 SEM equipped with a 4pi Analysis EDS system. A standard Be-window detector having a resolution of 158 eV was used. The beam voltage was adjusted depending on the x-ray lines of interest. The samples were mounted on an aluminum holder with a carbon coating.

D. Determination of boron and hydrogen concentration - Boron and hydrogen concentration was measured using a PGNAA gamma ray spectroscopy analysis at the MURR. The samples were placed in a neutron flux of approximately $1 x 10^8$ neutrons/cm^2-s. Prompt gamma ray emission is measured with a high purity germanium (n-type) detector. Both H and B can be distinguished by their unique gamma ray energy, 477.6 keV and 2223.3 keV respectively. A standard sample was used to determine the concentration of B and H in the samples.

E. Raman analysis - Raman analysis was performed in the Institute of Physics, Russian Academy of Science, Russia.

DISCUSSION

Optical inspection on the treated samples reveals an increase of the surface roughness of the samples. Blue, dark and metallic like spots were observed under the optical microscope at a magnification of 200x. The spots were located randomly on the samples. Based on the optical inspection, it is conclude that diffusion of the impurities was not homogeneous. As can be seen in figure 2 the color of treated diamond, original white to light gray, changed to a dark gray with some dark bluish coloration. It was found that as the boron concentration increased the sample became darker in color.

Figure 2. Treated diamond samples and untreated sample (right).

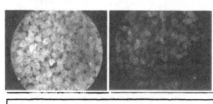

Figure 3. Optical micrograph, 20X. Untreated (left) and boron treated samples (right). Blue and dark spots colors were founded in some crystals.

Figure 3 shows the micrograph from treated and untreated diamond powder. It was found that some of the particles were colored and some not. Figure 4 shows different SEM micrographs of the samples. Micrograph a. show the surface of an untreated diamond, micrograph b – c shows the treated samples under different magnification. The analysis also demonstrates an increase of roughness in the samples after treatment. This observation agreed with the observation from optical inspection. It was also evident that the concentrations of impurities were not homogeneous through out the sample.

a. untreated   b. boron treated   c.

Figure 4 (a-c). SEM micrograph of natural diamond (a.) and treated (b-c) samples under different magnification.

Table 1. Experimental matrix used for the study of diffusion of Boron in the Diamond powder.

	TIME hrs	TEMP C°	BIAS DC	CURRENT amp	Boron ppm	Hydrogen ppm
P-OA	0	0	0	0	0	1207
P-01	4	950	150	~2X10^-6	120657	1524
P-03	6	950	150	~1	13,819	3368
P-04	16	950	150	~2X10^-6	31498	763
P-05	6	950	150	~1	118888	1000
P-07	20	950	150	~2X10^-6	57977	20

Table 1 shows the experimental matrix used to study the diffusion of boron in the diamond powder. Boron and hydrogen concentration was measured using Proton-Gamma

Neutron Activation Analysis. The values were compared to standard samples at the Missouri Research Reactor (MURR).

Figure 5a shows the experimental results from treated samples under different treatment times. A linear relationship was found between treatment time and boron concentration. The samples in Figure 5a correspond to samples P-03, P-4, P-07 in the experimental matrix of table 1. Those samples exhibit low conductivity during the experiment.

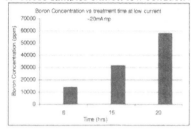
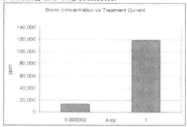

Figure 5a-5b. Comparison of boron concentration versus treatment time on samples treated with low current (left) and under high current (right).

The experimental relationship of boron concentration in those samples that exhibit high conductivity versus those samples that exhibit low conductivity is shown in Figure 5b. It was found that the boron concentration of those samples that exhibit high conductivity during the treatment were 3 to 4 times higher than the samples that did not exhibit high conductivity during the experiment. Samples treated with high current exhibit high conductivity and have a higher boron concentration. Figure 6 shows the comparison of boron content versus hydrogen content in the treated samples. It was found experimentally that the hydrogen concentration in the samples is lower when the boron concentration increases in the same sample. It is assume that boron and hydrogen compete in the diffusion process and for the interstitial position in the diamond lattice.

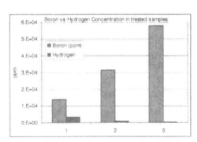
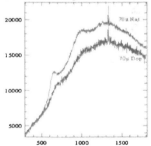

6a. B and H concentration on the BDD   6b. Raman analysis of the ND and BDD

Figure 6a Boron concentration (blue) versus hydrogen concentration (red) on different treated samples. 6b Raman analysis of the untreated sample(red) and treated sample(blue

CONCLUSIONS

Using the FEDOA technique we were successfully able to diffuse boron into natural diamond powders of 60-80 micron size. Transmitted light and SEM micrograph analysis confirmed an increase in surface roughness on treated samples. It was observed that a change in coloration in the diamond powder took place. It was also observed that the color darkened as the concentration of boron increased in the samples. Presence of impurities such as Mg and Si were also found by EDS analysis. It is assume the Mg and Si were present as impurities in the boron powder and in the system during the treatment process. Based on the experimental results relatively low temperatures were effective for FEDOA (750-950 C). It was found that after an elapsed diffusion time that the treated powder began to exhibit conductive properties. It was observed that the voltage in the power supply dropped from 150 V to ~17 volts and the current increased to ~2 amp. A direct relationship between the temperature and the current was indicated by a red-glow from the sample. It is known that diffusion of some impurities such as boron [5-7] can lead to conductivity in diamond. A possible transition between diamond to other phases of carbon can be responsible for the high conductivity found in samples. Those samples that exhibit the high conductivity are those with the higher boron concentration. It also appears that the current drag effect that enhances diffusion identified by Alexey Spitsyn et. al. [8,9] for silicon carbide also occurs in diamond. I-V measurement could be reproduced on the samples even after several consecutive cleaning of the samples suggesting that the conductivity was from the diamond samples. Electrical properties and I-V measurement of the powder will be presented in future work.

Special attention should be given to the hydrogen environment treatment. Hydrogen will diffuse into the diamond and react with the boron atoms. This can lead to the formation of boron hydrogen complexes inside the diamond. Another important step is to determine how and where the impurities are located in the diamond matrix. It is unknown if the impurities are diffused as an atom or as an ion. Also it is unknown if the impurities take interstitial or substitutional position in the diamond matrix. Further analysis including electro-chemical and optical characterization are going to be made on the samples. Some of the next steps will be the treatment of the boron-dope samples using hydrogen and deuterium plasma to study the absorption/desorption characteristics of the doped diamond powder.

REFERENCES

1. G. Popovici, T. Sung, M. Prelas, Forced Diffusion in Diamond: A Review, Journal of Chemical Vapor Deposition, Vol. 3, pages 115-132, October 1994.
2. M.A Prelas, Popovoci Galina, Sung,T. "Forced diffusion of impurities in natural diamond and polycrystalline diamond films", J. Appl. Phys v77 5625-5629 1995
3. M.A Prelas, Popovoci Galina, Sung,T. ,"Properties of diffused diamond films with n-type conductivity", Diamond and Related Material, v4 877 – 881 1995
4. Sung, T, M.A Prelas, "Doping Diamond by Forced Diffusion", dissertation , Univ. of Missouri-Columbia, May 1996
5. A.Mendez, M. Prelas, T. Ghosh, L. Ross, "Chromium diffusion on Natural Diamond", ANS Summer Meeting, June 5-9, 2005, San Diego, Ca.
6. A.Mendez, M. Prelas, T. Ghosh, L. Ross, "Study of Simultaneous Diffusion of Impurities into Natural Diamond Crystals" MSGC Annual Meting, April,2005
7. Prelas, Mark Popovoci, Galina, Bigelow, Louis., Handbook of industrial diamonds and diamond films, Mercel Decker 1998.
8. A.B. Spitsyn, M.A. Prelas, Robert V. Tompson, T.K. Ghosh, "Impurity Removal from 6-H SIC using reversed Field Enhanced Diffusion with Optical Activation," Journal of Wide Band-Gap Materials Volume 10/October 2002, pages 149-160 (Release date Sept. 2004).
9. A.B. Spitsyn, M.A. Prelas, Robert V. Tompson, T.K. Ghosh, "Impurity Removal from 4-H SIC using reversed Field Enhanced Diffusion with Optical Activation," Journal of Wide Band-Gap Materials Volume 10/October 2002, pages 89-98 (Release date Sept. 2004).

Mater. Res. Soc. Symp. Proc. Vol. 929 © 2006 Materials Research Society   0929-II05-06

The Corrosion Behavior of Nickel-Base Austenitic Alloys for Nuclear Hydrogen Generation

Rama S. Koripelli, Joydeep Pal, Ajit K. Roy
Mechanical Engineering, UNLV, Las Vegas, NV, 89119

ABSTRACT

Three nickel-base austenitic alloys, namely Alloy C-22, Alloy C-276 and Waspaloy have been tested for evaluation of their corrosion resistance in an acidic solution at ambient and elevated temperatures. The results of stress corrosion cracking studies indicate that none of these materials did exhibit any failure at constant load. The cracking susceptibility determined by the slow strain rate technique was gradually enhanced at higher temperatures showing reduced ductility and true failure stress. The critical potentials determined by the polarization technique, became more active (negative) with increasing temperature. The fractograpic evaluations by scanning electron microscopy (SEM) revealed ductile failure in Alloys C-22 and C-276. However, Waspaloy showed brittle failure at elevated temperature.

INTRODUCTION

The United States Department of Energy (USDOE) is currently exploring hydrogen generation, based on a thermochemical process known as sulfur-iodine (S-I) cycle. This process involves the utilization of heat from nuclear power plant to generate hydrogen by the formation and decomposition of sulfuric acid (H_2SO_4) and hydrogen iodide (HI), as illustrated in Figures 1 and 2 respectively. During this process the maximum temperature associated with decomposition of H_2SO_4 will be approximately 850°C. Thus, the structural materials to be used in the hydrogen generation plant must possess superior corrosion resistance due to the presence of H_2SO_4 at elevated temperatures.

Based on a literature review, three nickel-base alloys namely Alloy C-22, Alloy C-276 and Waspaloy have been identified as candidate structural material for heat exchanger applications in nuclear hydrogen generation. The susceptibility of all three alloys to stress corrosion cracking (SCC) and localized corrosion has been determined using different state-of-the-art techniques. Further, fractographic evaluations of the broken specimens have been determined by SEM. This paper presents the results of SCC, localized corrosion and fractogrphic evaluations of all three alloys.

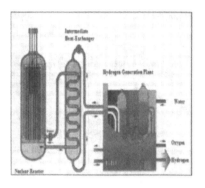

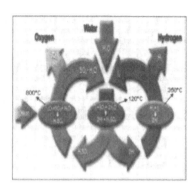

Figure 1. Hydrogen Generation Plant   **Figure 2.** S-I Cycle

EXPERIMENTAL PROCEDURES

Alloy C-22, Alloy C-276, and Waspaloy were procured from a vendor in a solution-annealed condition. The chemical composition and the metallurgical microstructures of all three alloys are given in Table I and Figure 3, respectively. Smooth and notched cylindrical specimens were machined from round bars of these alloys in such a way that the gage section was parallel to the longitudinal rolling direction. The stress concentration factor due to the presence of a notch in these cylindrical specimens was approximately 1.45 [1]. The susceptibility of these alloys to SCC was determined under both constant load and slow-strain-rate (SSR) conditions. A strain rate of 3.3×10^{-6} sec^{-1} was used in the SSR testing. The experimental details of these types of testing have been included in numerous reports [2].

A limited number of SCC test was also performed using self-loaded specimens such as C-ring and U-bend. These specimens were loaded according to the ASTM Designations G 38 [3] and G 30 [4] respectively. The susceptibility of these materials to localized corrosion was determined by the cyclic potentiodynamic polarization (CPP) technique, which has been described elsewhere [5]. The CPP tests were performed using a three-electrode technique, consisting of test specimen as the working electrode, two graphite rods as counter electrodes and silver/silver chloride (Ag/AgCl) as the reference electrode. The general corrosion rates of the tested materials were determined by using coupons. The extent and morphology of failure of the tested cylindrical specimens were determined by SEM. The metallurgical microstructures were evaluated by optical microscopy. For SCC and localized corrosion studies, an aqueous solution containing sodium iodide (NaI) and H_2SO_4 was used. However, coupons, C-ring and U-bend specimens were tested in an autoclave using an aqueous solution containing H_2SO_4 alone. The compositions of the testing solutions are given in Table II.

Table I. Chemical Composition of Materials Tested (wt %)

Materials	Elements																
	Ni	Cr	Mo	C	Fe	Co	W	Si	Mn	V	P	S	B	Zr	Al	Ti	Cu
Alloy C-276	Bal	14.5 to 16.5	15.0 to 17.0	0.01 (max)	4.0 to 7.0	2.5 (max)	3.0 to 4.5	0.08 (max)	1.0 (max)	0.35 (max)	0.02 (max)	0.01 (max)	-	-	-	-	-
Alloy C-22	Bal	20.0 to 22.5	12.5 to 14.5	0.01 (max)	2.0 to 6.0	2.5 (max)	2.5 to 3.5	0.08 (max)	0.5 (max)	0.35 (max)	-	-	-	-	-	-	-
Waspaloy	Bal	18.0 to 21.0	3.5 to 5.0	0.02 to 0.10	2.0 (max)	12.0 to 15.0	-	0.75 (max)	1.0 (max)	-	0.03 (max)	0.03 (max)	0.003 to 0.01	0.02 to 0.12	1.20 to 1.60	2.75 to 3.25	0.5 (max)

a) Alloy C-22   b) Alloy C-276

c) Waspaloy

Figure 3. Optical Micrographs showing Microstructures

Table II. Composition of Testing Environments (gram/liter)

Solution (pH)	NaI	H_2SO_4
Acidic (1.0)	10.0	Added to adjust the desired pH
Acidic (1.0)	-	Added with water to get the desired pH

RESULTS AND DISCUSSION

The results of SCC testing involving smooth cylindrical specimens at constant-load indicate that none of these tested materials exhibited any failure at applied stresses, equivalent to 95% of the materials' room temperature yield strength (YS) values. These results may suggest that the threshold stress for cracking in an aqueous solution containing NaI and H_2SO_4 may lie in the vicinity of 0.95YS of these materials. With respect to the cracking susceptibility of the notched specimens, the threshold load below which no failure was observed was in the vicinity of 70% of their yielding loads. The overall data obtained in the constant-load SCC testing are given in Table III.

The results of SCC testing using smooth specimens, performed under a SSR condition, are illustrated in Figures 4 through 6 for Alloy C-22, Alloy C-276 and Waspaloy, respectively in the form of an engineering stress versus strain (s-e) diagram. These figures also show the effect of temperature and environment on the s-e diagrams. An examination of these diagrams reveals that the failure strain was reduced to some extent at 90°C, suggesting that the cracking susceptibility was enhanced in the test environment at elevated temperature. The s-e diagrams obtained using notched specimens are illustrated in Figures 7 through 9, showing significantly reduced failure strain in the same testing environment at comparable temperatures. The reduced failure strain is an identification of increased cracking susceptibility due to the stress concentration effect associated with the presence of a notch. The magnitude of ductility in terms of percent elongation (%El), percent reduction in area (%RA), true failure stress (σ_f) and time to failure (TTF) determined from the s-e diagrams. As expected, all four parameters were significantly reduced due to the presence of a notch.

Table III. Constant-Load SCC Results

Material	Specimen Type	Environment, pH and Temperature	%Applied Stress/ Stress (ksi)	Failure/No Failure
Alloy C-22	Smooth Tensile		95% YS/50	No Failure
	Notch Tensile		70% YL/92	No Failure
Alloy C-276	Smooth Tensile	S-I,	95% YS/50	No Failure
	Notch Tensile	pH=1	50% YL/64	No Failure
	Notch Tensile	90 °C	70% YL/89	No Failure
	Smooth Tensile		95% YS/95	No Failure
Waspaloy	Smooth Tensile		95% YS/112	No Failure
	Notch Tensile		50% YL/135	No Failure

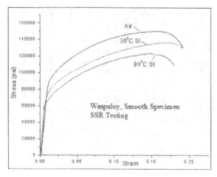

Figure 4. s-e Diagram, Smooth Specimen   **Figure 5.** s-e Diagram, Smooth Specimen

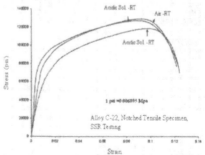

Figure 6. s-e Diagram, Smooth Specimen   **Figure 7.** s-e Diagram, Notched Specimen

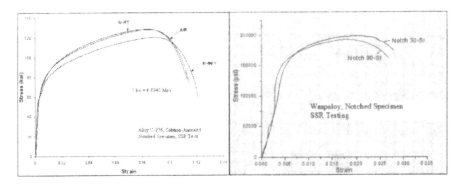

Figure 8. s-e Diagram, Notched Specimen   **Figure 9.** s-e Diagram, Notched Specimen

A typical CPP diagram for nickel-base alloys in the similar test solution is illustrated in Figure 10 showing forward and reverse scan. It appears that the CPP diagram for all three tested materials exhibited a negative hysterisis loop indicating resistance to localized attack. The formation of negative hysterisis loop may be the result of electrolysis leading to the disassociation of water into oxygen and hydrogen at high current density, thus forming new oxide layer on the surface. The CPP data also indicate that the magnitude of critical potentials became more active (negative) at higher temperatures. The variation of corrosion potential (E_{corr}) and pitting potential (E_{pit}) for all three alloys are given in table IV. None of the polarized specimens showed any localized attack.

The results of immersion testing inside an autoclave indicate that neither C-ring nor U-bend specimens did exhibit any cracking even after 56 days of exposure in acidic solution. However, the weight-loss data indicate that all three alloys showed insignificant dissolution up to 28 days followed by accelerated weight-loss after 56 days. The variation of weight-loss with exposure time is illustrated in Figure 11, showing the maximum weight-loss with Waspaloy. In view of these results, Alloy C-22 and Alloy C-276 may be the superior materials compared to Waspaloy in terms of general corrosion.

The fractographic valuations of the broken cylindrical specimens by SEM revealed dimpled microstructures for Alloys C-22 and C-276 at ambient and elevated temperature indicating ductile failure. However, for Waspaloy, some intergranular brittle failure was observed in the 90°C acidic solution. The SEM micrographs of Alloy C-22, Alloy C-276, and Waspaloy used in SSR testing are illustrated in Figure 12.

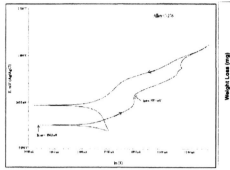 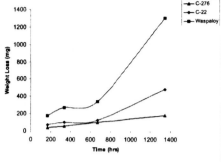

Figure 10. CPP Diagram, Alloy C-276   **Figure 11.** Autoclave Test Results, 150°C

Table IV. CPP Test Results

Temperature (°C)	Alloy C-22		Alloy C-276		Waspaloy	
	Average E_{corr} (mV)	Average E_{pit} (mV)	Average E_{corr} (mV)	Average E_{pit} (mV)	Average E_{corr} (mV)	Average E_{pit} (mV)
30°C	301	623	327	651	286	650
60°C	274	575	250	595	275	580
90°C	245	470	225	495	246	510

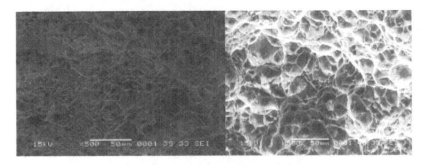

a) Alloy C-22, S-I solution, 500X   b) Alloy C-276, 90°C, S-I solution, 500X

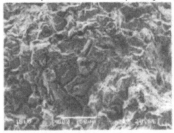

c) Waspaloy, 90°C, S-I solution, 500X
Figure 12. SEM Micrographs, SSR Testing

SUMMARY AND CONCLUSIONS

The cracking susceptibility of three nickel-base alloys has been determined by constant load and SSR techniques. The localized corrosion behavior was determined by the CPP technique. SEM evaluated the fracture morphology. The significant results are summarized below.

➢ No failure was observed at constant load. However, the cracking susceptibility was increased at elevated temperature in terms of reduced ductility and true failure stress, as determined by SSR testing.
➢ Both E_{corr} and E_{pit} became more active with increasing temperature.
➢ Ductile failure characterized by dimples was seen in Alloy C-22 and Alloy C-276. However, Waspaloy showed intergranular brittle failure at 90°C.

ACKNOWLEDGMENTS

Financial support by USDOE is thankfully acknowledged.

REFERENCES

1. March, J.L., Ruprecht, W. J., and Reed, George, "*Machining of Notched Tension Specimens*" ASTM Bulletin, ASTBA, Am.Soc.Testing Mats., No.244, 1960, pp.52-55.

2. ASTM Designation G129-00, *"Standard Practice for Slow Strain Rate Testing to Evaluate the Susceptibility of Metallic Materials to Environmentally Assisted Cracking"* American Society for Testing and Materials (ASTM) International.
3. ASTM Designation G38-01, *"Standard Practice for Making and Using C-Ring Stress-Corrosion Test Specimens"* American Society for Testing and Materials (ASTM) International.
4. ASTM Designation G30-97, *"Standard Practice for Making and Using U-Bend Stress-Corrosion Test Specimens"* American Society for Testing and Materials (ASTM) International.
5. ASTM Designation: G61-78, *"Standard Practice for Conducting Cyclic Potentiodynamic Polarization Measurements for Localized Corrosion"*, American Society for Testing and Materials (ASTM) International.

Mater. Res. Soc. Symp. Proc. Vol. 929 © 2006 Materials Research Society   0929-II05-08

Solar Effects on Tensile and Optical Properties of
Hubble Space Telescope Silver-Teflon Insulation

Kim K. de Groh[1], Joyce A. Dever[1], Aaron Snyder[1], Sharon Kaminski[2], Catherine E. McCarthy[2], Allison L. Rapoport[2], and Rochelle N. Rucker[2]

[1]Electro-Physics Branch, NASA Glenn Research Center, 21000 Brookpark Rd., M.S. 309-2, Cleveland, OH, 44135

[2]Hathaway Brown School, 19600 North Park Blvd., Shaker Heights, OH, 44122

ABSTRACT

A section of the retrieved Hubble Space Telescope (HST) solar array drive arm (SADA) multilayer insulation (MLI), which experienced 8.25 years of space exposure, was analyzed for environmental durability of the top layer of silver-Teflon fluorinated ethylene propylene (Ag-FEP). Because the SADA MLI had solar and anti-solar facing surfaces and was exposed to the space environment for a long duration, it provided a unique opportunity to study solar effects on the environmental degradation of Ag-FEP, a commonly used spacecraft thermal control material. Data obtained included tensile properties, solar absorptance, surface morphology and chemistry. The solar facing surface was found to be extremely embrittled and contained numerous through-thickness cracks. Tensile testing indicated that the solar facing surface lost 60% of its mechanical strength and 90% of its elasticity while the anti-solar facing surface had ductility similar to pristine FEP. The solar absorptance of both the solar facing surface (0.155 ± 0.032) and the anti-solar facing surface (0.208 ± 0.012) were found to be greater than pristine Ag-FEP (0.074). Solar facing and anti-solar facing surfaces were microscopically textured, and locations of isolated contamination were present on the anti-solar surface resulting in increased localized texturing. Yet, the overall texture was significantly more pronounced on the solar facing surface indicating a synergistic effect of combined solar exposure and increased heating with atomic oxygen erosion. The results indicate a very strong dependence of degradation, particularly embrittlement, upon solar exposure with orbital thermal cycling having a significant effect.

INTRODUCTION

The Hubble Space Telescope was launched on April 25, 1990 into low Earth orbit as the first mission of NASA's Great Observatories program. The HST was designed to be serviced on-orbit to upgrade scientific capabilities. In December 1993, during the first servicing mission (SM1) the original solar arrays (SA-I) were replaced with a second set of arrays (SA-II) containing bi-stem thermal shields to eliminate a thermal induced jitter problem. In March of 2002, after 8.25 years of space exposure, the SA-II was replaced with a third set of arrays (SA-III) during the fourth servicing mission (SM3B), and the SA-II was brought back to Earth. A section of the retrieved SA-II solar array drive arm (SADA) multilayer insulation (MLI) was provided to NASA Glenn Research Center so that environmental durability analyses of the top layer of silver-Teflon® fluorinated ethylene propylene (Ag-FEP) could be conducted.

The SADA MLI was wrapped completely around the SADA and therefore had solar and anti-solar facing surfaces (see Figure 1). The circular configuration of the insulation, along with the long-term space exposure, provided a unique opportunity to study solar radiation effects on the environmental degradation of Ag-FEP, a commonly used spacecraft thermal control material. The objective of this research was to characterize the degradation of retrieved HST SADA Ag-FEP after 8.25 years of space exposure with particular emphasis on the effect of solar exposure. Data obtained included tensile properties, solar absorptance, surface morphology and chemistry.

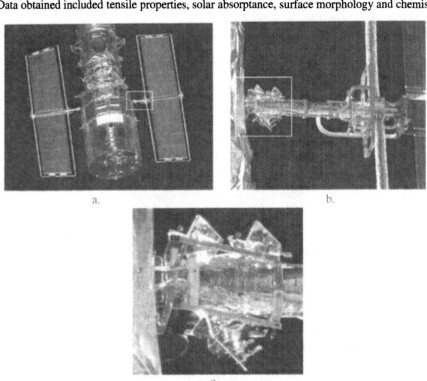

Figure 1. Hubble Space Telescope photographed in December 1999 during the third servicing mission (SM3A), viewed from the –V3 side: a). HST with SA-II, indicating the +V2 SADA, b). Close-up of the +V2 MLI covered SADA and c). Close-up of the in-board SADA section.

MATERIALS & ENVIRONMENTAL EXPOSURE

HST SM3B SA-II SADA MLI

The retrieved SA-II SADA MLI sample used in this study was provided by the European Space Agency (ESA) and is shown in Figure 2. The sample was 36.9 cm x 9.1 cm. The sample's top layer is approximately 10 mil (0.25 mm) thick, consisting of a space-exposed 5 mil (127 μm) Teflon® FEP layer coated on the backside with vapor deposited silver and Inconel (the Ag layer is 1500 Å thick and the Inconel layer is 275 Å thick). The FEP/Ag/Inconel layer is adhered with an acrylic adhesive (40 μm thick 966 acrylic) to a fiberglass cloth impregnated with polytetrafluoroethylene (PTFE). The MLI has 16 layers of double-sided aluminized-Kapton® (50 μm thick) separated by Dacron® netting and a bottom layer of PTFE-impregnated fiberglass cloth. All analyses were obtained from the top layer of insulation (FEP/Ag/ Inconel/adhesive/ scrim), referred to in this report as Ag-FEP.

This sample was cut from the +V2 SADA MLI section closest to the body of the telescope, or the in-board section (see Figures 1b & 1c). The V2 axis passes through the SADAs and the +V2 solar array is shown on the right in Figure 1a. For this study, 0° has been defined as the direct solar-facing surface (indicated as a line in Figure 2) and 180° is the anti-solar facing surface, with 90° and 270° being solar grazing surfaces. The exact position on the MLI sample that was directly solar facing was not known. Therefore, the 0° position was chosen based on the center of the discolored solar-facing region (described below). Pristine 5 mil (127 μm) thick Ag-FEP (without the adhesive or scrim) was used for obtaining pristine solar absorptance data and five pristine 5 mil (127 μm) thick FEP (non-metallized) samples were used for obtaining pristine tensile properties.

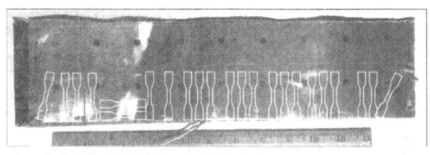

Figure 2. The HST SADA MLI sample (the line indicates the solar facing position (0°)).

Environmental Exposure

The estimated environmental exposure conditions for the SA-II SADA Ag-FEP, installed during SM1 and retrieved during SM3B, are given in Table 1, showing comparison to exposure conditions from the time of HST's deployment in April 1990 to each HST servicing mission and estimations of conditions through December 2010. Data from references 1 to 3 were used as the basis for estimating the environment exposure conditions. Environmental exposure conditions for the HST SA-I retrieved during SM1 were described in references 1 and 2. Environmental exposure conditions for SM1 through December 2010 were described in reference 3. It should be noted that cabling and other components covered areas of the SADA MLI insulation, primarily on solar grazing surfaces, as shown in Figures 1b and 1c. Also, Bay J, one of several small bays positioned on the anti-solar side of the telescope, is covered with aluminized-FEP MLI and could provide scattered atomic oxygen arrival to the anti-solar in-board SADA in addition to sweeping atomic oxygen arrival. The in-board anti-solar SADA might also receive somewhat reduced electron, proton and albedo UV radiation exposure due to the asymmetry of the telescope body around the in-board SADA (i.e. no bay is present on the solar side).

+V2 in-board SADA

Figure 3. Close-up photograph of HST taken during SM3A from the anti-solar side of the telescope showing Bay J and the +V2 SADA.

The HST mission timeline was used to determine the number of orbits and, therefore, the number of thermal cycles experienced from deployment to SM2.[3] In order to calculate the number of thermal cycles for SM3A, SM3B and December 2010, it is assumed that the average HST orbital period is 96 minutes. The equivalent solar exposure hours (ESH) data in Table 1 assume that the average HST orbit provides 32% of time in shadow, based on the variation of time-in-shadow between approximately 27-37%.[3] The SA-I environmental conditions reported in reference 2 assumed 38% time in shadow. Based on the 27-38% reported variation in percent of orbital shadow, error on calculated ESH is estimated to be ±11%. Equivalent Earth albedo solar exposure hours are also shown in Table 1. These data assume that directly solar-facing SADA surfaces receive Earth albedo illumination equivalent to 2.3% of direct solar exposure, and Earth albedo incident on anti-solar surfaces is equivalent to approximately 32% of direct solar exposure.[2]

Data for x-ray fluence in Table 1 were based on x-ray flux data from the Geosynchronous Operational Environmental Satellites (GOES) for the time period of launch (1990) through SM2 (February 1997).[3] For SM2 through 2010, the x-ray fluence was estimated assuming an average 11-year solar cycle.[3] Because x-ray exposure is primarily due to solar events, it is assumed that only solar facing surfaces receive x-rays and that anti-solar surfaces receive no x-ray exposure. HST was launched during a time of solar maximum; therefore, x-ray flux was high in the early part of the mission and decreased as solar minimum was reached around 1997. The next solar maximum peak is assumed to have been around 2001. Electron and proton fluence from solar wind particles trapped in Earth's magnetic field have been calculated previously using NASA's proton and electron models, AP-8 and AE-8, respectively.[3] The data for SM3B were extrapolated from these prior data. Electron and proton fluence is assumed to be omni-directional, providing exposure to both solar and anti-solar facing surfaces.

Table 1. Environmental Exposure Conditions for HST Ag-FEP Surfaces

Exposure		Deploy to SM1 (SA-I)	Deploy to SM2	Deploy to SM3A	Deploy to SM3B	SM1 to SM3B (SA-II)	Deploy to Dec. 2010
Missions		STS-31 & STS-61	STS-31 & STS-82	STS-31 & STS-103	STS-31 & STS-109	STS-61 & STS-109	STS-31 & N/A
Mission Dates		April 1990 & Dec. 1993	April 1990 & Feb. 1997	April 1990 & Dec. 1999	April 1990 & March 2002	Dec. 1993 & March 2002	April 1990 & N/A
Exposure (Yrs)		3.6	6.8	9.7	11.9	8.25	20.7
Thermal Cycles (#)		19,700	37,100	52,800	64,800	45,100	~110,000
Temperature Range (°C)		BOL Solar facing: -82 to -20 & Anti-solar facing: -137 to -30 EOL Solar facing: -60 to +30 & Anti-solar facing: -118 to + 5 SA I SADA, estimated: -100 to >+100 [1]					
ESH, Solar-facing	Direct	~21,500	~40,600	~57,500	~70,600	~49.100	~123,000
	Albedo	~500	~930	~1,300	~1,600	~1.100	~2,800
	Total	~22,000	~41,530	~58,800	~72,200	~50.200	~126,000
ESH, Anti-solar (albedo only)		6,900	13,000	18,400	22,500	15.600	39,300
X-ray Fluence (J/m^2)	1-8 Å	222.6	252.4	302.2	382.6	160.0	699.6
	0.5-4 Å	14.7	16.0	19.1	24.8	10.1	47.2
Electron Fluence (#/cm^2) > 40 keV		1.39×10^{13}	1.95×10^{13}	2.74×10^{13}	3.47×10^{13}	2.08×10^{13}	6.04×10^{13}
Proton Fluence (#/cm^2) > 40 keV		7.96×10^{9}	1.95×10^{10}	2.77×10^{10}	3.35×10^{10}	2.55×10^{10}	5.86×10^{10}
Atomic Oxygen Fluence (atoms/cm^2)	Ram	9.6×10^{20}	1.0×10^{21}	1.8×10^{21} *	2.2×10^{21}	1.2×10^{21}	$3.4\text{-}3.5 \times 10^{21}$
	Solar	1.2×10^{20}	1.3×10^{20}	2.3×10^{20}	2.7×10^{20}	1.5×10^{20}	4.4×10^{20}
	Anti-solar	1.5×10^{20}	1.6×10^{20}	2.8×10^{20}	3.4×10^{20}	1.9×10^{20}	5.5×10^{20}

* In reference 3, atomic oxygen fluence was calculated assuming a mission date of June 2000. The fluence shown in Table 1 has been linearly interpolated for the actual mission date (December 1999).

Atomic oxygen ram fluence values for HST indicated in Table 1 were modeled based on the Mass Spectrometer Incoherent Scatter Model 86 (MSIS-86).[3] Using the ram fluence values for each HST servicing mission, atomic oxygen fluences for solar-facing and anti-solar facing surfaces were obtained using equations 1 and 2, respectively.[3]

$$Fluence_{solar\text{-}facing} \approx Fluence_{ram} \times 0.2528 \times 0.5 \qquad (1)$$

$$Fluence_{anti\text{-}solar} \approx Fluence_{ram} \times 0.3167 \times 0.5 \qquad (2)$$

The solar-facing and anti-solar atomic oxygen fluence values are provided in Table 1. Equations (1) and (2) use flux ratios for solar-facing/ram and anti-solar/ram determined for

International Space Station surfaces.[4] Equations (1) and (2) also assume that the SADA is close enough to the body of HST to have a view of only one half the angles seen by unobstructed surfaces, resulting in half of the fluence that would be incident on unobstructed surfaces (hence the 0.5 factor).

Temperature range estimates for the solar facing and anti-solar facing SA-II drive arm MLI were provided by Goddard Space Flight Center.[5] These values were computed using the HST Detailed SA-III Thermal Model using a representative geometric shape for SA-II. Four cases were run to get temperature extremes:

1) *Cold attitude for the sun side: beginning-of-life (BOL) optical properties, cold environmental fluxes (this provides the minimum temperature on the sun side)*
2) *Hot attitude for the sun side: end-of-life (EOL) optical properties, hot environmental fluxes (this provides the maximum temperature on the sun side)*
3) *Cold attitude for shade side: BOL optical properties, cold environmental fluxes (this provides the minimum temperature on the shade side)*
4) *Hot attitude for shade side: EOL optical properties, hot environmental fluxes (this provides the maximum temperature on the shade side)*

Beginning-of-life solar absorptance (α_s) and thermal emittance (ε) of 0.08 and 0.81, respectively were used. End-of-life α_s and ε values of 0.187 (0.155 ± 0.032) and 0.81, respectively, were used (see Solar Absorptance results section). For the hot attitude case the solar flux was 1,419 W/m^2, the albedo was 0.35 and the Earth's infrared (IR) was 265 W/m^2. For the cold attitude case the solar flux was 1,286 W/m^2, the albedo was 0.25 and the Earth's IR was 208 W/m^2. There was no conduction considered between blanket nodes (i.e. the surface of the blanket). However, a blanket emittance to the mast of 0.03 was included, along with conduction in the mast in both the radial and lateral directions. Using the listed BOL and EOL optical properties, the resulting temperature estimates are lower than those provided by European Space Agency (-100 to +100 °C) in reference 1.

EXPERIMENTAL PROCEDURES

Optical and Electron Microscopy

An Olympus SZH microscope operated with a Canon EOS D30 digital camera was used to document various features of the sample at magnifications on the order of 10X to 100X. Scanning electron microscope images were obtained using a Hitachi S-4700 field emission scanning electron microscope (FESEM) operated at an accelerating voltage of 6 kV. Energy dispersive spectroscopy (EDS) was conducted using an EDAX CDU Leap Detector system.

Tensile Properties

Tensile properties (ultimate tensile strength (UTS) and elongation at failure) were obtained using a DDL Inc. Model 200Q Electromechanical Test System. Twenty-three samples were punched out of the MLI sample (see Figure 2) using a die fabricated to the specifications defined in the American Society for Testing and Materials (ASTM) Standard D-638 for Type V tensile specimens.[6] The "dog-bone" tensile samples had a length of 63.4 mm and width of 9.52 mm, with a 7.62 mm long and 3.18 mm wide neck region. Care was taken to avoid cracks and impact sites and hence tensile samples were not obtained at precise angular intervals. The initial grip separation distance was set for 25.4 mm and testing was conducted at a speed of 1.072 mm/sec. Stress versus strain graphs were produced for each of the tested samples. For computation of UTS the thickness was set to 0.010" (0.254 mm) to take into account the total thickness of the sample (FEP/Ag/Inconel/adhesive/scrim), even though the FEP thickness is only 0.005" (0.127 mm). Thus, the UTS values for the HST SADA Ag-FEP are not comparable with pristine FEP, which does not have the added adhesive and scrim layer.

The HST SADA Ag-FEP stress versus strain graphs had three distinct steps indicating breakage of three distinct layers. The scrim layer broke with very little elongation, followed by breakage of the FEP layer, and finally breakage of the ductile adhesive layer. Therefore, the UTS and elongation at failure of the FEP layer was determined based on the location of the second step (break of the FEP) on the stress vs. strain graphs. See Figure 4 for an example.

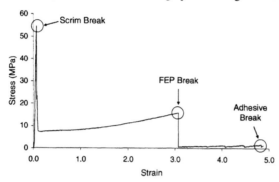

Figure 4. Stress-strain curve for Tensile Sample 5 (295.6°) showing three break points.

Solar Absorptance

An AZ Technology Laboratory Portable SpectroReflectometer 200 (LPSR-200) was used to measure total hemispherical spectral reflectance in the 250-2500 nm wavelength range. The LPSR-200 automatically integrates the reflectance data to calculate solar absorptance in accordance with standard methods described in reference 7. Since the sample was embrittled, care was taken to keep the sample flat during data acquisition. Data were originally taken at 9 different solar angle locations. Solar absorptance data were then taken at an additional 52 positions along the top section of the sample (the section above the dog bone samples shown in Figure 2). Three readings were taken at each of the 52 locations and averaged. It should be noted that the position of the absorptance values (with respect to solar angle) may have some error as the sample was not sectioned, but was positioned underneath the LPSR-200. Hence, exact locations were more difficult to determine than for the tensile samples.

Thermal emittance (ϵ) typically does not change due to environmental exposure unless substantial erosion has occurred. Hence an emittance of 0.81, reported in reference 8, was used for both the pristine and HST SADA 5 mil (0.127 μm) thick Ag-FEP.

RESULTS & DISCUSSION

Optical Microscopy

The HST SADA sample had distinct areas with varying discoloration. In general, the solar-facing region had a non-uniform hazy-white, almost milky appearance, and contained numerous through-thickness cracks. Two of these through-thickness cracks are shown in Figure 5. Some areas in the solar facing region were discolored brown. There was no observable darkening long the cracks, hence no significant oxidation of the underlying Ag layer was observed. The solar grazing surfaces had clearer appearances, similar to pristine Ag-FEP, and did not contain cracks. The anti-solar facing side was uniformly hazy-white in appearance, but did not contain any observable cracks.

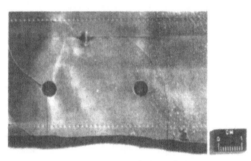

Figure 5. Close-up of the solar facing side of the SADA MLI showing through-thickness cracks and discolored regions.

Tensile Properties

Polar plots of UTS and elongation at failure versus solar angle are provided in Figure 6. As can be seen, the solar facing surfaces (0°) are severely embrittled as witnessed by significant decreases in both the UTS and elongation. The UTS and elongation at failure of pristine FEP were 27.4 ± 1.2 MPa and $285.3 \pm 15.5\%$, respectively. Although direct comparison of the UTS values cannot be made with pristine FEP because of the thickness differences, large differences in the UTS are observed for the SADA MLI as a function of solar angle, as shown in Figure 6a. The average UTS of the anti-solar facing surface is 13.2 ± 0.5 MPa (average of 8 data points between 133° and 223°). The solar grazing surfaces have very similar average UTS values: 12.4 ± 1.5 MPa for 4 data points between 64° and 122° and 12.4 ± 1.5 MPa for 5 data points between 238° and 296°. These values are slightly lower than the average anti-solar facing UTS. The solar facing surface has greatly decreased average UTS of 5.4 ± 1.1 MPa (averaged from 6 data points between 31° and 322°). This is a 60% decrease compared to the anti-solar surface UTS.

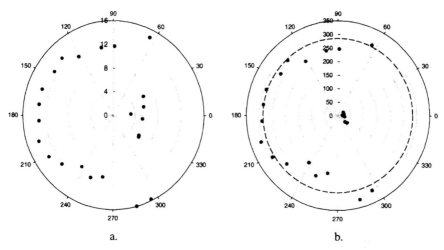

Figure 6. Polar plots of the HST SADA Ag-FEP tensile data: a). UTS (MPa) vs. solar angle, and b). Elongation at failure (%) vs. solar angle (the dashed line is for pristine FEP).

Elongation at failure is not dependent on sample thickness; therefore, the elongation at failure of the HST SADA Ag-FEP can be directly compared with that of pristine FEP, as shown in the polar plot in Figure 6b. By comparing the SADA Ag-FEP elongation to that of pristine FEP (shown as a dashed line), it is apparent that the anti-solar facing surfaces have elongations (284.8 ± 14.3%) essentially the same as pristine FEP (285.3 ± 15.5%), while the solar grazing surfaces have somewhat decreased elongation as compared with pristine FEP (average elongations of 254.0 ± 25.2% and 258.3 ± 52.7% for data centered on 90° and 270°, respectively). The solar facing surface is extremely embrittled with a greatly decreased elongation averaging only 29.8 ± 9.3%. This is a 90% decrease compared to pristine FEP. Severe embrittlement with through thickness cracking of significantly darkened solar facing sections of the MLI had already occurred after 6 years of space exposure, as photo-documented during SM3A.

A HST MLI Failure Review Board (FRB) was assembled after large cracks were observed in the HST light shield MLI during SM2. Studies conducted by the FRB on retrieved HST FEP insulation and ground testing of pristine FEP indicated that thermal cycling with deep-layer damage from electron and proton radiation are necessary to cause FEP embrittlement with the propagation of cracks along stress concentrations, and damage increases with the combined total dose of electrons, protons, ultraviolet and x-ray radiation along with thermal cycling.[9] The SA-II SADA Ag-FEP results appear somewhat inconsistent with the FRB findings: as electron and proton radiation is omni-directional one would expect the anti-solar facing SADA MLI to be somewhat embrittled. However, the results of these analyses indicate otherwise.

Studies conducted by de Groh et. al. and de Groh and Martin have shown the impact of temperature on the degradation of irradiated FEP. These studies provide evidence that the dominant mechanism of degradation of FEP in the space environment is irradiation induced chain scission and that heating allows chain mobility resulting in increased crystallization and therefore increased embrittlement.[10,11] A dynamic thermal model, which computes temperature throughout the orbit and includes heat conduction in the circumferential direction, was developed at Glenn Research Center to determine the variation of temperature versus solar angle for the HST SADA Ag-FEP. An example of the temperature variation is provided in Figure 7. This plot was computed for an HST orbit station positioned between the Earth and Sun. At this orbit position, direct solar heating occurs (1,400 W/m^2 = one solar unit) on the solar side, and albedo heating and Earth's IR heating on the anti-solar side. However, to take into account reflection from the body of the telescope onto the solar side of the SADA, an arbitrarily chosen direct solar heating of 1.5 solar units (2,100 W/m^2) was used. The albedo heating factor was 0.35 solar units (490 W/m^2) and the IR heating was 0.19 solar units (265 W/m^2), equivalent to an Earth temperature of -12 °C. Internal heating was not considered in this case. Two plots are shown in Figure 7: the lower temperature plot (open circle symbol) was computed using pristine optical properties ($\alpha_s = 0.08$ and $\varepsilon = 0.81$), thus representing the BOL temperature range. The higher temperature plot (solid circle symbol) was computed using optical properties of the solar facing side of the retrieved insulation ($\alpha_s = 0.187$ and $\varepsilon = 0.81$), thus representing the EOL temperature range. An alpha of 0.187 (0.155 + 0.032) was used from the values provided in the Solar Absorptance section below. These plots show how the temperature varies from the solar facing to anti-solar facing surfaces, and how increased solar absorptance greatly increases the temperature. The temperature variation between the solar and anti-solar facing surfaces, combined with the presence of solar radiation on the solar facing surface, appears to play an important role in the variation in embrittlement versus solar angle of the Ag-FEP.

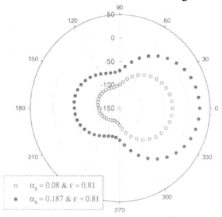

Figure 7. Temperature for the SA-II SADA Ag-FEP as a function of solar angle.

Using the temperature values computed with the Glenn dynamic thermal model for the EOL optical properties ($\alpha_s = 0.187$ and $\varepsilon = 0.81$), a graph was made of the elongation at failure versus temperature. This graph is provided in Figure 8 and shows a couple different trends. The first is that surfaces heated to the highest temperatures (over 0 °C) were the most dramatically degraded, while the surfaces heated to lower temperatures were still ductile. Yet this plot also shows a trend for decreasing elongation, and hence embrittlement, for temperatures dropping below -50 °C. Transition temperatures occur in FEP over two broad temperature ranges depending on hexafluoropropylene (HFP) content: between 83 °C and 150 °C [10] and between

-93 °C and -13 °C [12]. Perhaps the downward trend shown with decreasing temperature in this chart is due to effects of being below the cold transition temperature (during radiation exposure). Although this might be true, the effect is not nearly as deteriorative, with respect to loss of mechanical properties, as the higher temperature exposure.

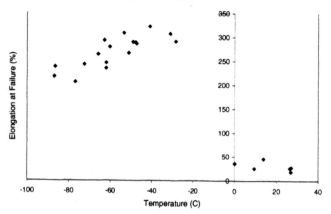

Figure 8. Elongation at failure of the SADA Ag-FEP versus temperature for EOL optical properties based on the Glenn dynamic thermal model.

Solar Absorptance

A polar plot of the solar absorptance versus solar angle is provided in Figure 9. The solar absorptance for pristine Ag-FEP was 0.074 and is plotted as a dashed line. As can be seen, the solar absorptance for all 61 locations on the HST SADA MLI is higher than the pristine Ag-FEP solar absorptance. The polar plot shows greater increases in absorptance for the solar and anti-solar surfaces and smaller increases at the solar grazing angles (areas likely shielded by cabling and equipment).

The data was averaged based on 60° angular views centered on 0, 90, 180 and 270°. The solar facing surfaces had an average absorptance of 0.155 ± 0.032 (average of 12 data points between 26.4° and 337.5°) with a large amount of scatter. The scatter in the data corresponds to the variation in haziness and brown discoloration found on the solar facing surfaces. The average solar absorptance values for the two solar grazing surfaces were similar: an average of 0.129 ± 0.027 between 249.6° and 287.9° (average of 10 data points) and 0.134 ± 0.014 between 77° and 114.2° (average of 9 data points). Thus, the overall average for the solar grazing surfaces was 0.131 ± 0.021 (average of 19 data points). Surprisingly, the anti-solar surfaces had high solar absorptance of 0.208 ± 0.012 (average of 11 data points between 151.4° and 201°). The scatter in this data was much smaller than for the solar facing data, which is consistent with the more uniformly hazy appearance of the anti-solar surface.

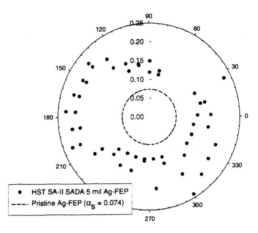

Figure 9. Polar plot of the solar absorptance versus solar angle for the HST SADA Ag-FEP.

Scanning electron microscopy revealed differences in the texture morphology of the SADA Ag-FEP as a function of solar angle, as shown in Figure 10, which appear to correspond to the differences in solar absorptance. The solar facing surface had a wavy rill-like textured morphology, which is attributed to sweeping atomic oxygen erosion (Figure 10a). The solar grazing surface was similar to pristine FEP in appearance, smoother with manufacturing lines visible (Figure 10b), which is consistent with the location receiving very little atomic oxygen exposure. Yet at very high magnifications, some areas of the solar grazing surfaces are slightly textured as compared to pristine FEP indicating that this particular area was not shielded by cabling or equipment. The anti-solar side was non-uniformly textured (Figure 10c). It had an overall texture that was somewhat rougher than the solar grazing surface, but there were non-uniform isolated locations of greater texture where the surface appeared to have been protected from erosion. This surface received slightly higher atomic oxygen exposure than the solar facing surface, so one might expect a similar texture.

Energy dispersive spectroscopy analyses indicated the presence of F, C and O on the "typical texture" areas for all surfaces. Analyses conducted on the tops of the "protected" buttes, such as shown in Figure 10, indicate contaminants such as Na and Si, which appear to have provided isolated protection against atomic oxygen erosion.

It has been debated whether there is a synergistic effect of solar exposure on the atomic oxygen erosion of FEP. This study appears to support that theory, as the overall texture is much smoother on the anti-solar side. Perhaps there is a joint effect of simultaneous solar radiation exposure combined with higher temperature exposure during atomic oxygen exposure resulting in greater texturing on the solar facing surface. One potential complicating factor in this analysis is the contribution of scattered atomic oxygen (and reduced radiation) on this surface from Bay J. Substantial modeling would need to be conducted to determine the extent of atomic oxygen scattering and its possible effect on the erosion depth and texture, which is beyond the scope of this research.

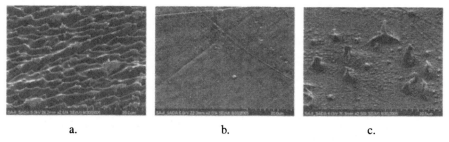

a.   b.   c.

Figure 10. Scanning electron microscopy images of SADA Ag-FEP surfaces: a). Solar facing texture (2.5 kX), b). Solar grazing texture (2 kX), and c). Anti-solar facing texture (2.5 kX).

NASA Glenn Research Center has designed two Materials International Space Station Experiment (MISSE) space exposure experiments, which together will address solar radiation synergistic effects on the atomic oxygen erosion of FEP and other fluoropolymers. One of these experiments has already been flown and retrieved and the other is currently being exposed to the space environment on the exterior of the International Space Station.

CONCLUSIONS

A section of retrieved HST SA-II SADA thermal insulation that received 8.25 years of space exposure was provided to NASA Glenn Research Center so that environmental durability analyses of the top layer of Ag-FEP could be conducted. The solar facing surface of the insulation was found to be non-uniformly hazy in appearance and embrittled with numerous through-thickness cracks. Tensile testing indicated that the solar facing surface lost 60% of its mechanical strength and 90% of its elasticity. The anti-solar facing surface was also found to be hazy in appearance, but tensile testing indicated that the anti-solar facing surface had ductility similar to pristine FEP, which was unexpected. Asymmetry of the telescope near the in-board section of the SADA may have resulted in reduced electron, proton and albedo UV radiation of the anti-solar surface. A thermal model was developed to determine temperature versus solar angle for the insulation. Plotting elongation at failure versus temperature indicated a trend of increasing embrittlement for decreasing temperatures below -50°C, appearing to correspond to a cold transition temperature, but this effect was not nearly as deteriorative as the solar facing higher temperature exposures. These results indicate a very strong dependence of embrittlement on solar exposure, and the effect of temperature appears to play an important role.

The solar absorptance values of the solar facing surface (0.155 ± 0.032) and the anti-solar facing surface (0.208 ± 0.012) were found to have significantly increased compared to the solar absorptance of pristine Ag-FEP (0.074). Scanning electron microscopy and corresponding EDS indicated that both the solar facing and anti-solar facing surfaces were microscopically textured, affecting the solar absorptance, and the anti-solar side had locations of isolated Si contamination that contributed to increased localized texturing. Yet, the overall texture was significantly more pronounced on the solar facing surface, even though it received less atomic oxygen, indicating a synergistic effect of combined solar exposure and increased heating with atomic oxygen erosion. But it should be kept in mind that the anti-solar surface may have received scattered atomic oxygen from Bay J in addition to sweeping atomic oxygen, which would modify the erosion texture. These results provide valuable information on space environmental degradation of Ag-FEP, particularly with respect to solar radiation and temperature effects on embrittlement.

ACKNOWLEDGEMENTS

The authors would like to thank the European Space Agency and the Hubble Project Office at Goddard Space Flight Center, and in particular Ben Reed, for providing the HST insulation sample for these analyses. They would like to thank Janet Hodermarsky of Ohio Aerospace Institute for conducting the initial photo-documentation and the tensile tests. The authors also greatly appreciate the thermal modeling support provided by Joshua Abel and Elisabeth Abel of Lockheed Martin at Goddard Space Flight Center. Finally, they would like to thank David Bretz of the Image Science & Analysis Group for helping to provide numerous high resolution images of the Hubble Space Telescope during various servicing missions.

REFERENCES

1. M. Van Eesbeek, F. Levadou and A. Milintchouk, "Investigation on FEP from PDM and Harness from HST-SA1", Proceedings of the Hubble Space Telescope Solar-Array Workshop, Noordwijk, the Netherlands, May 30-31, 1995, ESA WPP-77, 403-416 (1995).
2. G. Drolshagen, "Definition of the Space Environment for the HST Solar-Array 1", Proceedings of the Hubble Space Telescope Solar-Array Workshop, Noordwijk, the Netherlands, May 30-31, 1995, ESA WPP-77, 53-65 (1995).
3. J. A. Dever, K. K. de Groh, B. A. Banks, J. A. Townsend, J. L. Barth, S. Thomson, T. Gregory, W. Savage, *High Performance Polymers*, **12**, 125-139 (2000).
4. Rockwell International Corporation 1990, Rocketdyne Division, Specification RC 1800 Review C, p. 18.
5. Personal communication with Joshua Abel, Hubble Space Telescope CHAMP Program Lead Thermal Engineer, Lockheed Martin and Elisabeth Abel, Thermal Engineer, Lockheed Martin.
6. American Society for Testing and Materials ASTM D 638-95, "Standard Test Method for Tensile Properties of Plastics," 1995.
7. ASTM E 903-82, "Standard Test Method for Solar Absorptance, Reflectance, and Transmittance of Materials Using Integrating Spheres," American Society for Testing and Materials, 1982, Re-approved 1992.
8. John H. Henninger, "Solar Absorptance and Thermal Emittance of Some Common Spacecraft Thermal Control Coatings," NASA RP 1121, 1984.
9. J. A. Townsend, P. A. Hansen, J. A. Dever, K. K. de Groh, B. A. Banks, L. Wang and C. He, *High Performance Polymers*, **11**, 81-99 (1999).
10. K. K. de Groh, J. R. Gaier, R. L. Hall, M. P. Espe, D. R. Cato, J. K. Sutter and D. A. Scheiman, D A 2000 *High Performance Polymers*, **12**, 83-104 (2000).
11. K. K. de Groh and M. Martin, Journal of Spacecraft & Rockets, Vol. 41, No. 3, 366-372 (2004).
12. R. P. Reed, R. E. Schramm and A. F. Clark, *Cryogenics* **13**, 67–82 (1973).

Mater. Res. Soc. Symp. Proc. Vol. 929 © 2006 Materials Research Society   0929-II04-03

Surface Modification of Glassy Polymeric Carbon by Glow Discharge

Volha Abidzina[1], I. Tereshko[1], I. Elkin[2], R.L. Zimmerman[3], I. Muntele[3], C. Muntele[3], R.A. Minamisawa[3], B. Chhay[3], and D. Ila[3]

[1]Belarusian-Russian University, Prospect Mira 43, Mogilev, 212005, Belarus

[2]KAMA VT' Research and Production Enterprise, Karl Libknecht Str. 3a, Mogilev, 212000, Belarus

[3]Center for Irradiation of Materials, Alabama A&M University, Normal, AL, 35762-1447

ABSTRACT

We have studied the effects of low energy ion beam produced by glow discharge on glassy polymeric carbon (GPC) produced at pyrolysis temperature ranging from 200C to 1000C. The pristine and irradiated GPC were studied using FTIR, Raman Spectrometry indicating changes at the surface of the GPC samples.

INTRODUCTION

Glassy Polymeric Carbon (GPC) is made from cured phenolic resins by heat-treatment in inert environment. During the carbonization, increasing the heat treatment temperature, chemical changes take place that eliminate noncarbon elements by diffusion while aromatic rings condense to form graphitic lamellar layers. The final structure of GPC is composed of these graphitic layers in random arrangement with uncorrected pores between them and the material appears dark, hard and vitreous [1]. GPC is an impermeable isotropic carbon material and it can be fabricated with high purity. Due to its chemical inertness and biocompatibility, GPC is useful for medical application such as heart valves and other prosthetic devises [2].

It was shown that MeV ion bombardment significantly enhances the surface roughness and porosity of the partially and fully precursor phenolic resins [3]. MeV bombardment by ions such as silicon, carbon, oxygen and gold increases the hardness of GPC [4]. The increase in hardness is greater for carbon bombardment and it is a result of increased linkage between the chain of the polymeric material, which is due to the recombination of broken bonds or the generation of free radicals by the process of bombardment with energetic ions.

GPC is also used for the harsh environment of space as well as for protective coating against extreme environments such as high temperature, highly ionizing radiation, as well as corrosive environment. Therefore we conducted this investigation on the influence of low energy ion irradiation.

EXPERIMENTAL METHODS

GPC samples were prepared at the pyrolysis temperatures of 200, 550 and 1000°C [5,6]. After curing at 60°C the resin is pyrolyzed at low temperature rates to avoid changing shape or disruption due to production of volatile gases. Heat treatment to 550°C produces a conducting material due to hydrogen release and conjugation of the aromatic rings forming graphene planes in random arrangement. Even at 650°C pyrolysis temperature, the material has high level of available pores. Further heat treatment to 1000°C shows reducing permeability as the graphitic planes aggregate random grapheme planes.

Before the irradiation in glow-discharge plasma some of GPC samples were coated with Au or Ti. GPC samples were placed into a plasma generator and were exposed to the low energy bombardment by the ions produced from the residual gases. The energy of ions depends on the voltage in the plasma generator and did not exceed 3 keV. The current in the plasma generator was maintained at 40-50 mA. Barometric pressure of residual gases in the plasma generator chamber was 5.3 Pa. Irradiated dose was $2 \cdot 10^{17}$ ion·cm^{-2}.

GPC samples were analyzed with Fourier transform infrared (FTIR) and Raman spectrometry. The two regions of interest (D- and G-line) in the Raman analysis were investigated. G-line is attributed to graphene structure formation in the material and D-line is attributed to amorphous structure, 'disorder' in the material, as well as the vibration of the bond at the edge of the graphitic ribbons.

RESULTS AND DISCUSSION

Fig. 1(a) shows FTIR spectra for the GPC samples made at 200°C that were coated either with Au or Ti or irradiated for 20 minutes in comparison with a pristine sample, i.e. the one that did not exposed any irradiation. An increase in C=C bonds is observed for the samples coated with Au or Ti.

Fig. 1 (b) shows FTIR spectra for the GPC samples made at 1000°C before and after coating of Ti or Au or after irradiation.

The spectra show a decrease in C-H bonds in comparison with a pristine sample that may correspond to the graphene formation on the surface.

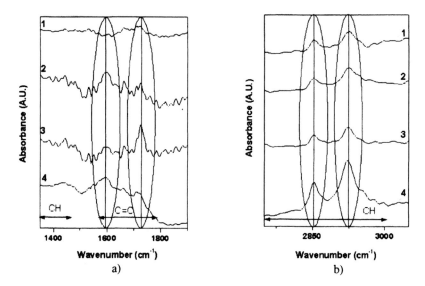

Figure 1. FTIR spectra for GPC samples.
a) prepared at 200°C, 1 – GPC sample coated with gold and then irradiated for 20 min, 2 – GPC sample coated with Ti, 3 - GPC sample coated with Au , 4 – pristine GPC sample
b) prepared at 1000°C, 1 – GPC sample coated with Ti, 2 – GPC sample coated with Au, 3 – GPC sample coated with gold and then irradiated for 20 min, 4 – pristine sample.

Using Raman spectroscopy the ratio of D/G bands for pristine and irradiated GPC surfaces were compared (Fig. 2).

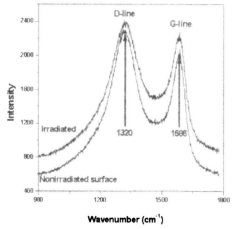

Figure 2. Raman spectrum for GPC sample coated with Au and irradiated for 20 minutes

The results indicate an increase in D/G ratio for the GPC samples made at 1000°C coated with Ti followed by irradiation for 20 minutes and a decrease in the D/G ratio for the samples which were coated with Au and for the ones that were irradiated for 20 minutes (Table 1).

Table 1. D/G ration for GPC prepared at 1000°C.

Type of GPS samples	Nonirradiated surface	Irradiated surface
GPS coated with Ti	1.15	1.18
GPS coated with Au	1.14	1.11
GPS coated with Au and irradiated for 20 minutes	1.13	1.08

GPC samples made at 550°C were coated by either Au or Ti or irradiated for 20 and 40 minutes. Comparing the Raman spectra for both surfaces we have observed that D/G ratio stays constant except for the sample that was irradiated for 40 minutes.

Table 2. D/G ration for GPC prepared at 550°C.

Type of GPS samples	Nonirradiated surface	Irradiated surface
GPS coated with Ti	0.95	0.95
GPS coated with Au	0.95	0.95
GPS irradiated for 20 minutes	0.95	0.95
GPS irradiated for 40 minutes	0.95	0.97

The Raman spectra of the GPC samples made at 200°C show the presence of D-line and a weak G-line. Any sputtering or irradiation influences decrease D/G ratio. Fig. 3 shows Raman spectra for the GPC sample that was exposed to the low-energy ion irradiation for 30 minutes.

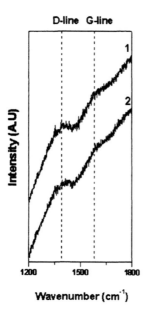

Figure 3: Raman spectra for GPC samples made at 200°C and irradiated for 30 minutes: 1 – nonirradiated surface, 2 – irradiated surface.

CONCLUSIONS

FTIR spectra of GPS samples made at 1000°C which were coated either with Ti or Au or irradiated in glow-discharge plasma indicate graphene formation on the surface. Raman spectra of these samples show changes in D/G ratio depending on the type of coating and the duration of the exposure to low energy ion irradiation.

Neither coatings nor low energy irradiation of GPC produced at 550C cause any observable change in the D/G ratio.

The low temperature pyrolyzed of resin at 200C coated either with Au or Ti resulted to an increased in C=C bonds whereas the irradiation or coating by Au or Ti resulted to a decrease in D/G ratio.

ACKNOWLEDGMENT

Research sponsored by the Center for Irradiation of Materials, Alabama A&M University and in part by the AAMURI Center for Advanced Propulsion Materials under the contract number NAG8-1933 from NASA, and by National Science Foundation under Grant No. EPS-0447675.

REFERENCES

1. G. M. Jenkins and K. Kawamura, Polymeric Carbons-Carbons Fiber, Cambridge University Press, 1976
2. G.M. Jenkins, D. Ila, H. Maleki, Mat. Res. Soc. Symp. Proc., 394 (1995) 181
3. R. Zimmerman, D. Ila, C. Muntele, M. Rodrigues, D.B. Poker, D. Hensley, Nuclear Instruments and Methods in Physics Research, B 191 (2002) 825
4. M.G. Rodrigues, N.C. da Cruz, E.C. Rangel, R.L. Zimmerman, D. Ila, D.B. Poker, D.K. Hensley, Nuclear Instruments and Methods in Physics Research, B 191 (2002) 524
5. H. Maleki, L.R. Holland, G.M. Jenkins, R.L. Zimmerman, Journal of Material Research, 11-9 (1996) 2368
6. H. Maleki, D. Ila, G.M. Jenkins, R.L. Zimmerman, A.L. Evelyn, Material Research Society Symposium Proceedings, 371 (1995) 443

Mater. Res. Soc. Symp. Proc. Vol. 929 © 2006 Materials Research Society 0929-II04-04

Nanostructural Evolution of Au on Silica Surfaces Exposed to Low Energy Ions

Volha Abidzina[1], I. Tereshko[1], I. Elkin[2], R.L. Zimmerman[3], S. Budak[3], B. Zheng[3], C. Muntele[3], and D. Ila[3]

[1]Belarusian-Russian University, Prospect Mira 43, Mogilev, 212005, Belarus
[2]KAMA VT Research and Production Enterprise, Karl Libknecht Str. 3a, Mogilev, 212000, Belarus
[3]Center for Irradiation of Materials, Alabama A&M University, Normal, AL, 35762-1447

ABSTRACT

We studied the effects of the low energy ions to induce nucleation of nanoscale crystals on and near surface of silica nano-layer containing low concentrations of Au. Suprasil substrates were coated with thin layer of gold followed by low-energy ion irradiation in glow discharge plasma. The formation of nanoscale crystals due to low energy ion irradiation was then studied using RBS and optical absorption spectrometry.

INTRODUCTION

During the last decade, metallic ion implantation and thermal annealing have been used to change the linear and the non-linear optical properties near the surface of silica glass [1-8]. An attractive property of ion implantation is that ions can be focused in a well-defined space in an optical device, to induce local changes in its linear and nonlinear properties.

The classical treatment of small spheres imbedded in an optical material of index n_o shows that an imposed electric field produces in each sphere a dipole moment proportional to the field and to the factor

$$\frac{\varepsilon - n^2_o}{\varepsilon + 2n^2_o}, \tag{1}$$

where ε is the dielectric constant of the material of the spheres. For spheres of conducting materials $\varepsilon = \varepsilon_1 + j\varepsilon_2$, where the real component ε_1 is negative and the imaginary component ε_2 is proportional to the conductivity which causes energy loss from a time varying electric field, such as that in visible light. Mie [8] derived the optical absorption coefficient α of a material with a volume fraction Q occupied by metal spheres whose radii are small compared with incident light of wavelength λ

$$\alpha = \frac{18\pi Q 2n^3_o}{\lambda} \cdot \frac{\varepsilon_2}{(\varepsilon_1 + 2n^2_o)^2 + \varepsilon^2_2} \quad (cm^{-1}) \tag{2}$$

A minimum occurs in the denominator of equation (2) when

$$\varepsilon_1(\lambda_p) + 2n^2_o = 0 \tag{3}$$

and causes a maximum absorption of light at a characteristic wavelength λ_p, the so called surface plasmon resonance [1, 2].

Doyle [9] showed that for spheres whose size less than the mean free path of the conduction electrons the plasmon resonance is broader not by conductivity, represented by ε_2, of the bulk material but by the radius r of the spheres. For the full width at half maximum $\Delta\lambda$ of the peak determined from an optical absorption measurement

$$r = \frac{v_f \lambda^2_p}{2\pi c \Delta\lambda},$$ (4)

where v_f is the electron velocity corresponding to the Fermi energy of the metal. λ_p depends on the substrate and the element implanted it in and $\Delta\lambda$ is related to the size of the nanoclusters. Using Mie's and Doyle's theories is in good correlation with our previous works obtained by TEM [10, 11].

Changes in the optical properties of insulating materials can be made either by MeV bombardment or by heat treatment. The aim of this paper is to investigate changes in the optical properties caused by low-energy ion irradiation in glow discharge plasma.

EXPERIMENTAL PROCEDURES

Silica substrates were coated by a thin film of gold and irradiated by low-energy ion using a glow-discharge residual gas plasma. The ion energy depended on the voltage in the plasma generator and did not exceed 3 keV. The current in the plasma generator was 40-50 mA.

Barometric pressure of residual gases in the plasma generator chamber was 5.3 Pa. The irradiation dose was maintained at $2 \cdot 10^{17}$ ions·cm^{-2}. The temperature in the chamber was controlled during the irradiation process and did not exceed 323 K while the irradiation time varied from 10 to 60 minutes.

Using optical absorption spectrometry and Rutherford Backscattering spectrometry (RBS) we studied the formation of nanoclusters after low-energy ion irradiation.

RESULTS AND DISCUSSION

Figure 1 shows RBS spectrum for the sample sputtered with gold and irradiated in glow discharge plasma for 30 minutes. The spectrum indicates the presence of gold on the surface as well as the contamination such as iron that might appear from the high-carbon steel cathode.

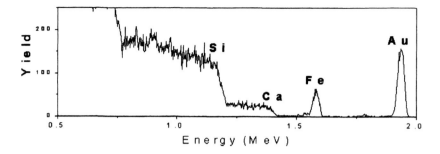

Figure 1. RBS spectrum for silica sample sputtered with gold and irradiated for 30 minutes

Figure 2 shows the values of permittivity (ε_1) versus wavelength using the CRC data tables for bulk metals [11]. The downward trend of ε_1, linear in wavelength, together with the slow variation of ε_2, leads us to expect quasi Lorentzian optical absorption peaks centered at a wavelength λ_p characteristic of the index of refraction of the host and of the permittivity of the metal. Using $n_o = 1.5$ for silica and the value of ε_1 from Fig.1, equation (3) predicts wavelength of 535 nm for gold and 340, 503 and 820 nm for iron.

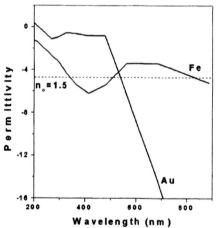

Figure 2. The permittivity for gold and iron

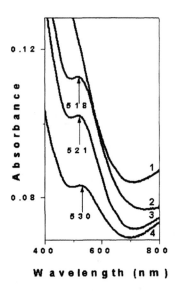

Figure 3. Optical absorption spectra: 1 – sample sputtered with gold, 2 – gold coated silica irradiated for 20 minutes, 3 - gold coated silica irradiated for 30 minutes, 4 - gold coated silica irradiated for 60 minutes

Figure 3 compares the optical absorption spectra for gold-coated silica irradiated in glow discharge plasma. The experimental optical absorption spectra for gold are in reasonable agreement with equation (2). We observed the appearance of peaks for the samples that were irradiated more that 20 minutes. In addition, increasing the time of irradiation we observed a slight shift in wavelength.

From equation (4), and the position and width of the optical absorption peak, we infer that increasing time of irradiation, radii of nanoclusters are increasing. For 20 minutes of the irradiation they correspond to 8.3 nm, for 30 minutes – 9.2 nm and for 60 minutes – 9.7 nm. The formation of the absorption band is due to the formation of nanoclusters of metals such as Au in Silica as per Mie Theory predicted it. The increase in the absorption base or the reduced transmission through the substrate is due to damage to the Silica substrate causing scattering of light. Generally the absorption or scattering by substrate is reduced if the substrate is annealed after ion beam processing.

CONCLUSIONS

Low energy ion bombardment, by glow discharge, of gold-coated silica causes formation of nanoscale gold particles as observed by optical absorption spectrometry. The longer exposure to the glow discharge provides an absorption band at 530 nm which is equal to the predicted value by Mie theory.

ACKNOWLEDGEMENTS

Research sponsored by the Center for Irradiation of Materials, Alabama A&M University and in part by the AAMURI Center for Advanced Propulsion Materials under the contract number NAG8-1933 from NASA, and by National Science Foundation under Grant No. EPS-0447675.

REFERENCES

1. G. W. Arnold, *J. Appl. Phys.* **46**, 4466 (1975)
2. G. W. Arnold and J. A. Bordes, *J. Appl. Phy.* **48**, 1488 (1977).
3. R. H. Magruder, R. A. Zuhr, D. H. Osborne, Jr., *Nucl. Inst. & Meth. in Phys. Res.* **B99**, 590 (1995).
4. R. F. Haglund, Jr., D. H. Osborne, Jr., R. H. Magruder, III, C. W. White, R. A., Zuhr, P. D. Townsend, D. E. Hole, and R. E. Leuchtner, *Mat. Res. Soc. Symp. Proc.* **Vol. 354**, 629 (1995).
5. C. W. White, D. S. Zhou, J. D. Budai, R. A. Zuhr, R. H. Magruder and D. H. Osborne, *Mat. Res. Soc. Symp. Proc.* **Vol. 316**, 499 (1994).
6. K. Fukumi, A. Chayahara, M. Adachi, K. Kadono, T. Sakaguchi, M. Miya, Y. Horino, N. Kitamura, J. Hayakawa, H. Yamashita, K. Fujii and M. Satou, *Mat. Res. Soc. Symp. Proc.* **Vol. 235**, 389 (1992).
7. D. Ila, Z. Wu, R. L. Zimmerman, S. Sarkisov, C. C. Smith, D. B. Poker, and D. K. Hensley, *Mat. Res. Soc. Symp. Proc.* **457**, 143 (1997).
8. G. Mie, *Ann. Physik* 25, 377 (1908).
9. W.T. Doyle, Phys. Rev. 111, 1067 (1958).
10. D. Ila, R.L. Zimmerman, C.I. Muntele, P. Thevenard, F. Orucevic, C.L. Santamaria, P.S. Guichard, S. Schiestel, C.A. Carosella, G.K. Hubler, D.B. Poker, D.K. Hensley, *Nucl. Inst. & Meth. in Phys. Res.* **B191**, 416 (2002).
11. D. Ila, E.K. Williams, R.L. Zimmerman, D.B. Poker, D.K. Hensley, *Nucl. Inst. & Meth. in Phys. Res.* **B166-167**, 845 (2000).
12. David R. Lide, Ed., Handbook of Chemistry and Physics 76, CRC Press, New York 1995, pp 12-126-133

Mater. Res. Soc. Symp. Proc. Vol. 929 © 2006 Materials Research Society   0929-II04-10

Effect of MeV Si Ion Bombardment on Thermoelectric Characteristics of Sequentially Deposited SiO$_2$/Au$_x$SiO$_{2(1-x)}$ Nanolayers

S. Budak[1], B. Zheng[1], C. Muntele[1], Z. Xiao[2], I. Muntele[1], B. Chhay[1], R. L. Zimmerman[1], R. L. Holland[1], and D. Ila[1]

[1]Physics/CIM, Alabama A&M University, 4900 Meridian Street P.O. Box 1447, Normal, AL, 35762

[2]Electrical Engineering, Alabama A&M University, 4900 Meridian Street, Normal, AL, 35762

ABSTRACT

We made an electro-cooling superlattice system consisting of 50 and 100 periodic nano-layers of SiO$_2$/Au$_x$SiO$_{2(1-x)}$ with Au layers deposited on both sides as metal contacts. The deposited multi-layer films have an alternating layers of 10 nm thick. The ultimate objective of this research is to tailor the Figure of Merit of layered structures used as thermoelectric generators. The superlattices were then bombarded by 5 MeV Si ions at three different fluences to form nano-cluster structures. The film thicknesses and stoichiometry were monitored by Rutherford backscattering spectrometry (RBS) before and after MeV ion bombardments. We measured the thermoelectric efficiency of the fabricated device before and after the MeV ion bombardments. We measured the cross plane thermal conductivity by the 3rd harmonic method, the cross plane Seebeck coefficient, and the electrical conductivity using the Van Der Pauw method before and after the 5 MeV Si bombardments.

INTRODUCTION

Thermoelectric materials are important due to the interest in their applications in thermoelectric power generation and microelectronic cooling [1]. Effective thermoelectric materials have a low thermal conductivity and a high electrical conductivity [2]. The performance of the thermoelectric materials and devices is shown by a dimensionless Figure of Merit, $ZT = S^2 \sigma T / \kappa$, where S is the Seebeck coefficient, σ is the electrical conductivity, T is the absolute temperature, and κ is the thermal conductivity [3]. ZT can be increased by increasing S, increasing σ, or decreasing κ.

In this study we report on the growth of a $SiO_2/Au_xSiO_{2(1-x)}$ Superlattice on the silica and silicon substrates using an ion-beam assisted deposition (IBAD) system, and high energy Si ion bombardments of the films for reducing thermal conductivity and increasing electrical conductivity.

EXPERIMENTAL

We have grown $SiO_2/Au_xSiO_{2(1-x)}$ superlattice nanolayer films on the silica and silicon substrates with the IBAD system. The multilayer films were sequentially deposited to have a periodic structure consisting of alternating SiO_2 and $Au_xSiO_{2(1-x)}$ layers. The two electron-gun evaporators for evaporating the two solids were turned on and off alternately to make multilayers. The thickness of the layers was controlled by an INFICON deposition monitor. The film geometry used in this study is shown in Fig.1. The geometry in Fig.1 shows two Au contacts on the top and bottom of the multilayers. These contacts were used in the Seebeck coefficient measurement system.

The electrical conductivity was measured by the Van der Pauw system and the thermal conductivity was measured by the 3ω technique. The thermal conductivity measurement was performed at a room temperature of $22\,^{\circ}C$. One can find detailed information about this technique elsewhere [4]. The 5 MeV Si ion bombardments were performed by the Pelletron ion beam accelerator at the Alabama A&M University's Center for Irradiation of Materials (AAMU-CIM). The energy of the bombarding Si ions was chosen by the SRIM simulation software.

The fluences used for the bombardment were $1x10^{13}\,ions/cm^2$, $5x10^{13}\,ions/cm^2$, and $1x10^{14}\,ions/cm^2$. The lowest thermal conductivity was found at the fluence of $1x10^{14}\,ions/cm^2$, and the highest electrical conductivity was found at the same fluence. The RUMP simulation software was used to analyze the Rutherford backscattering (RBS) spectra for the element concentrations in the films [1].

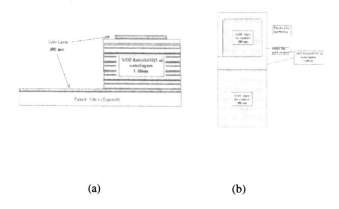

(a)   (b)

Fig.1. Schematic geometry of the sample (a) from the cross-section, and (b) from the top.

RESULTS AND DISCUSSION

Fig.2 shows the RBS spectra from 50 periodic nano-layers of $SiO_2/Au_xSiO_{2(1-x)}$ where the multilayer films were grown on glassy polymeric carbon (GPC) substrates. The sources of the RBS ions were helium and nitrogen. We need to make the multilayer films with less periodicity so that we could resolve and observe each layer clearly.

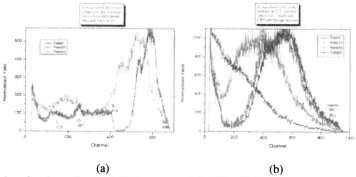

(a)   (b)

Fig. 2. Angular dependence of RBS spectra of $SiO2/Au_xSiO_{2(1-x)}$ superlattice on GPC substrate when the source was (a) helium and (b) nitrogen.

The data taken the 50 and 100 period of $SiO_2/Au_xSiO_{2(1-x)}$ superlattice nanolayer films were shown by Fig. 3-5. Fig.3a shows the Seebeck Coefficient Variation of the 50 periodic nano-layers of $SiO_2/Au_xSiO_{2(1-x)}$ superlattice. As the temperature increased, the Seebeck coefficient increased in the positive direction. Similar effect can be seen in Figure 3b for 100 periodic case.

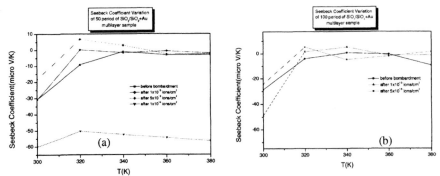

Fig.3. Seebeck Coefficient Variation of (a) 50 periodic and (b) 100 perodic nano-layers of $SiO_2/Au_xSiO_{2(1-x)}$ super lattice

Fig. 4 shows thermoelectric properties of 50 periodic nano-layers of $SiO_2/Au_xSiO_{2(1-x)}$ superlattice. Fig. 4a shows the Seebeck coefficient dependence on the fluences. As seen from the figure, the Seebeck coefficient started to increase at the initial fluence and decreased after $5x10^{13} ions/cm^2$. The Fig. 4b shows the square of the Seebeck coefficient change depending on the fluences of the bombardment. Fig.4c shows the electrical conductivity change depending on the fluences. This shows that the ion bombardment caused the increment in the electrical conductivity. Fig. 4d shows the thermal conductivity change depending on the fluences. The figure shows that the thermal conductivity decreases as the bombardment fluence increases. This is also another desired situation in thermoelectric materials and devices. The high energy ion bombardment can produce nanostructures and modify the property of thin films [5], resulting in lower thermal conductivity and higher electrical conductivity.

Fig.4e shows the Figure of Merit change depending on the ion bombardment. As seen from Fig.4e the Figure of Merit increases as the fluence increases. This is the solution to the major problem of thermoelectric materials. The ion bombardment showed that the Figure of Merit increases as the fluence of the bombardment increases when suitable fluences are chosen. The Figure of Merit for this sample increased from 0.66×10^{-4} to 2.52 at fluences of 0 and $1 \times 10^{14} \, ions/cm^2$, respectively.

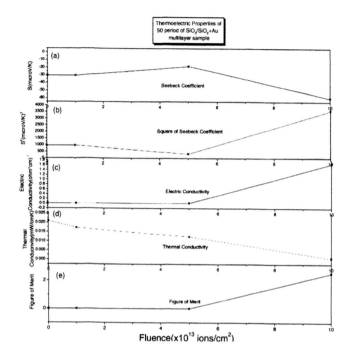

Fig.4. Thermoelectric Properties of 50 periodic nano-layers of $SiO_2/Au_xSiO_{2(1-x)}$ superlattice

Fig.5 shows the electrical conductivity of 100 periodic nano-layers of $SiO_2/Au_xSiO_{2(1-x)}$ superlattice. As seen from the figure the electrical conductivity increases when the fluence increases.

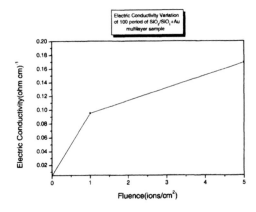

Fig.5. Electrical conductivity of 100 periodic nano-layers of $SiO_2/Au_xSiO_{2(1-x)}$ superlattice

References

1. Z. Xiao, R.L. Zimmerman, L.R. Holland, B. Zheng, C.I. Muntele, D. Ila, Nuclear Instruments and Methods in Physics Research B 242 (2006) 201–204

2. Brian C. Scales, 15 February 2002 vol 295 science

3. Z. Xiao, R.L. Zimmerman, L.R. Holland, B. Zheng, C.I. Muntele, D. Ila, Nuclear Instruments and Methods in Physics Research B 241 (2005) 568–572

4. L.R. Holland, J.Apl. Phys. 34 (1963) 2350

5. D. Ila, R.L. Zimmerman, C.I. Muntele, P. Thevenard, F.Orucevic, C.L. Santamaria, P.S. Guichard, S. Schiestel, C.A. Carosella, G.K. Hubler, D.B. Poker, D.K. Hensley, Nuclear Instruments and Methods in Physics Research B 191 (2002) 416

Mater. Res. Soc. Symp. Proc. Vol. 929 © 2006 Materials Research Society   0929-II04-19

Damage Effects of Ionizing Radiation in Polymer Film Electrets

Marco Aurélio Parada[1], Renato Amaral Minamisawa[1,2], Marcos Vasques Moreira[3], Adelaide de Almeida[1], Iulia Muntele[4], and Daryush Ila[2]

[1]1Departamento de Física e Matemática, Universidade de São Paulo, Avenida Bandeirantes 3900, Ribeirão Preto, São Paulo, 14040-901, Brazil

[2]2Center for Irradiation of Materials, Alabama A&M University, P.O.Box 1447, Normal AL, Huntsville, Alabama, 35762-1447

[3]Instituto de Radioterapia e Megavoltagem de Ribeirão Preto (IRMEV), Rua Sete de Setembro 1150, Ribeirão Preto, São Paulo, 14010-180, Brazil

[4]Physics, Alabama A&M University, 4900 Meridian Street, Normal, AL, 35762

ABSTRACT

Electret sensors or dosimeters can be used to quantify the ionizing radiation dose from charged particles or waves beams (α, β, -e, p etc... γ, X), and with appropriate converters, from fast and slow neutrons. The electret state is reached, by some insulating materials (electrical conductivity lower than 10^{-8} (Ω m)$^{-1}$), when once charged, the incorporated charge is quasi-permanent $\approx 10^9$ s. The charge densities are read (before and after the irradiations) and the radiation dose inferred from the difference between them. PFA (Tetrafluoroethylene-per-fluoromethoxyethylene) and FEP (Tetrafluoroethylene-hexa-fluoropropylene) damage mechanisms were studied bombarding these fluoropolymers with: 1 MeV protons at constant current and fluences from 1×10^{11} tower 1×10^{16} ions/cm^2, ^{60}Co gamma and X-rays, respectively of 1.25 and 0.106 MeV for absorbed doses of 0.5, 1.0, 8,0 and 100 Gy. The emission of chemical species was monitored with a Residual Gas Analyzer (RGA), during proton bombardment and techniques of Optical Absorption Photospectrometry (OAP), Fourier Transform Infrared (FTIR) and Micro-Raman spectroscopy were used to analyze the virgin, exposed and irradiated films.

INTRODUCTION

Fluorocarbons present high mechanical and electrical resistance, also are resistent to high temperatures. The combination of these characteristics is suitable for: aerospace engineering applications (once they resist damages from the ionizing radiation met in space) [1,2], replacement for some human body tissues, pharmaceutical packing and also as radiation dosimeters [3,4]. Electret dosimeters can be produced with FEP and PFA films, initially charged by corona process [4].After charging, their external electrical fields attract the ions resulting from the radiation interaction with the air, near to their surfaces, and these ions compensate charges initially incorporated. The compensated charge differential is proportional to the radiation dose and a linear behavior (calibration curve) of charge *versus* radiation absorbed dose can be obtained for the dosimetry of several types, geometry, energy and intensities radiation [3]. The film remaining charge can be measured, through induction process, with an appropriate reader system composed of a probe (selected according to the film geometry) and an electrometer [3,4]. Once fluoropolymers are viable for dosimetry, it is important to know how they behave under the conditions of several fluences of protons bombardment and photons absorbed dose.

MATERIALS AND METHODS

The polymers films used in all the experiments had 20x20 mm^2 area and 25 μm thick and all experimental conditions were kept constant during the measurements.

In order to characterize the PFA and FEP films, regard to the incorporated charge densities after the corona process, a planar probe was developed [6,7] to be used in the charge reader system (figure 1). For that, another fluoropolymer (ETFE - Ethylenetetrafluoroethylene), was also measured for more comparison in the ratio of carbon-fluorine fraction by weight, for each film chemical composition *versus* the measured charge densities.

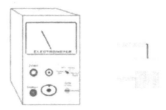

Figure 1. Electret charge density reader system with the planar probe (lateral and frontal views).

Table I. ETFE, FEP and PFA composition.

Film	Fraction by weight (w/w)			
	Carbon	Fluorine	Oxygen	Hydrogen
ETFE	0.375	0.571	----	0.031
FEP	0.174	0.826	----	----
PFA	0.217	0.687	0.096	

The films damage mechanisms, due to the proton and photon irradiation, were analysed with the techniques of OAP, for the ultraviolet-visible region, and FTIR. Additional information about PFA and FEP structures were obtained by Micro-Raman analyses [8].

PFA and FEP films were bombarded with proton of 1 MeV in the AAMU/CIM Pelletron at fluences of $1x10^{11}$, $1x10^{12}$, $1x10^{13}$, $1x10^{14}$, $1x10^{15}$ and $1x10^{16}$ protons/cm^2 (one film for each fluence) and the current was kept below 1 μA, to avoid excessive sample heating. Residual Gas

Analysis was conducted in real time (RGA/Stanford Research System/200) during the bombardment.

PFA and FEP were also irradiated with absorbed doses of 0.5, 1.0, 4.0, 8.0 and 100 Gy in the IRMEV Theratronics/Theratron 780C and Philips/Stabilipan RT200/250, respectively with gamma and X-rays at different dose rates of 0.659 Gy/min for gamma rays and 1.488 Gy/min for X-Ray. The samples were also analyzed with OAP, FTIR and Micro-Raman Spectroscopy.

RESULTS AND DISCUSSION

From the surface charge density results for each film, it was observed that the ETFE presents the highest charge density value, more near to PFA than to FEP. In order to evaluate surface charge storage capacities (related with their chemical structures) for each film, the carbon and fluorine fraction by weight (table I) ratio values, were plotted *versus* charge surface density. These results are presented in figure 2, where one can notice that the film with highest charge storage capacity has also the highest carbon-fluorine fraction by weight ratio.

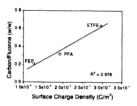

Figure 2. Carbon/Fluorine (W/W) *versus* Surface charge density for ETFE, PFA and FEP.

The FTIR and Micro-Raman results obtained with virgin polymer samples are presented in figure 3. From the results PFA and FEP structures are very similar, differing only by the presence of oxygen atoms in the PFA. The evidence of the carbon triple bond is explained by the presence of cross link chains in the undamaged polymer structures. There is no evidence of the carbon double bonds.

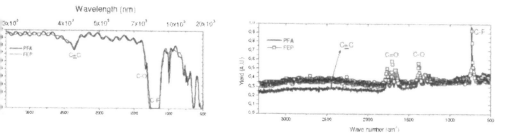

Figure 3. FTIR and Micro-Raman spectra from PFA and FEP virgin films.

In figure 4 and figure 5 are presented the RGA results for the gases emitted simultaneously with the films proton bombardment. The gases emission, during the bombardment, indicates modifications in the polymers due to broken bounds in the polymeric chains. CF_3 radicals account for the greater part of the partial pressure detected during the bombardment for both films, but other species were also emitted, such as CF, COF, CF_2, and CF_2CCF_3. In the same figures are presented the relative gas emissions, during the bombardment, where one can see that the gas emission slowly decreases with the increase of proton fluence. This fact may indicate that the accumulation of positive surface charge is sufficient to deflect part of the proton beam, decreasing the emission yield from the polymeric chains. More significantly, the decrease of CF_3 and CF_2CCF_3 species emission from the FEP, indicates that this film is extensively altered by the depletion of these species on the surface, when the fluence is highest.

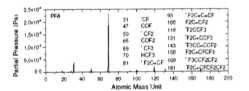 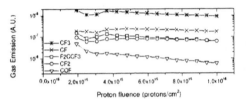

Figure 4. PFA RGA and Gas emission results *versus* proton fluence.

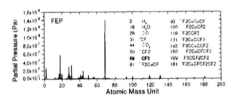 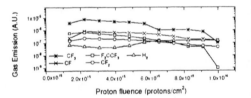

Figure 5. FEP RGA and Gas emission results *versus* proton fluence.

The results for Optical Absorption Photospectrometry are shown in the figure 6. In these figures are presented the spectra obtained from virgin and bombarded films at the fluences indicated. The spectra for the virgin films indicate that single carbon bounds are present in the polymer structure. After bombardment, one can note that the spectra are dislocated to the right, indicating the carbon double bounds formation.

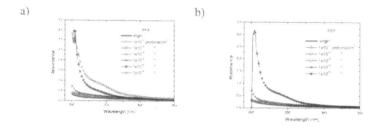

Figure 6. UV-Visible spectra of virgin and proton bombarded films. a) PFA with fluences of $1x10^{11}$ - $1x10^{16}$ protons/cm^2 and b) FEP with fluences of $1x10^{11}$ - $1x10^{15}$ protons/cm^2.

In figure 7 are presented the FTIR spectra for virgin and bombarded films at fluences of $1x10^{14}$ and $1x10^{16}$ protons/cm^2 for PFA polymer and at $1x10^{13}$ and $1x10^{15}$ protons/cm^2 for FEP polymer. The FEP FTIR and OAP spectra couldn't be obtained at $1x10^{16}$ ions/cm^2 due to damage produced in that bombarded film. The spectrum for the highest fluence confirms the carbon double bound formation in the polymeric chains.

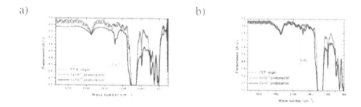

Figure 7. FTIR spectra for virgin and bombarded films. a) PFA with fluences of $1x10^{14}$ - $1x10^{16}$ protons/cm^2 and b) FEP with fluences of $1x10^{14}$ - $1x10^{16}$ protons/cm^2.

The OAP results for PFA and FEP virgin and charged films are presented in figure 8a), where one can observe that there is no significant absorbance difference between the virgin and charged films. Virgin films present higher absorbance values, what could be interpreted as refraction index changes, due to the charges injected in their surfaces. In figure 8b) the OAP for PFA and FEP irradiated films are presented, as an example for the absorbed dose of 8.0 Gy. The absorbance values inferred for the same photons energies (1.250 and 0.106 MeV) for others absorbed dose values (0.5; 1.0; 4.0 Gy) are similar and they do not present significant differences from those of the virgin and charged films. No new peaks were noticed in the irradiated films spectra, evidence of no new chemical bond formation in [200 – 400 nm].

a)　　　　　　　　　　　　　　　b)

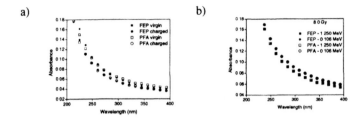

Figure 8. a) OAP results for virgin and charged PFA and FEP films, b) OAP from PFA and FEP with gamma and X-rays photons at an absorbed dose of 8 Gy.

The FTIR results for a) PFA and b) FEP films are presented in figure 9. In this figure the spectra for the virgin (transmission), 0.106 and 1.250 MeV irradiated films are shown for several absorbed dose with 8.0 Gy for the ATR. The PFA spectrum for 100 Gy X-ray irradiated film does not show significant damage. The peaks shown for virgin and irradiated films (up to 8.0 Gy) are in the same region for transmission and ATR spectra. It is observed no relevant peak formation in the region from 2500 to 750 cm^{-1}, but spectra from films irradiated with 100 Gy X-rays showed evidence of relevant material chemical damage. Similar results for the FEP films are presented in part b).

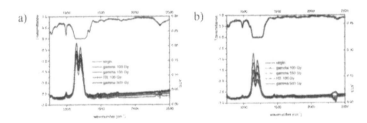

Figure 9. FTIR and ATR results for FEP irradiated films with photons of 1.25 and 0.106 MeV for different absorbed doses. a) For PFA film and b) for FEP film.

In figure 10 are shown the Micro-Raman results for PFA and FEP virgin and irradiated (8.0 Gy) films with the same photon energies, as those used in the FTIR analysis. A new peak formation for FEP irradiated with photons of 0.106 MeV indicates evidence of a chemical change provoked by the irradiation. For FEP, relatively small damage occurred only for the 8.0 Gy absorbed dose. Results for the same film bombarded with 1.0 MeV protons [9] showed that damage that can compromise its use as a radiation dosimeter occurs only for fluences higher than 10^{14} protons/cm^2 and absorbed doses higher than 8.0 Gy.

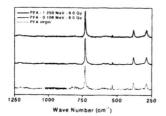 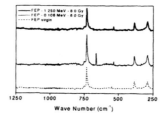

Figure 10. Micro-Raman results for PFA and FEP irradiated films with photons of 1.25 and 0.106 MeV at an absorbed dose of 8.0 Gy.

CONCLUSIONS

For protons bombardment the RGA results indicate that several chemical species are emitted from the film surfaces, being the $^•CF_3$ the one preferentially emitted, once it is a radical in the polymeric chain. In the OAP and FTIR spectra one can note the formation of new bands, indicating that these techniques are adequate to evaluate the film damages, due to the proton bombardment.

Considering the use of these films as proton dosimeters, the film damage which could compromise the applications in dosimetry occur only for fluence greater than 10^{15} protons/cm^2, when after the bombardment the films become fragile

For ^{60}Co photons irradiation for all absorbed doses, evidence of chemical damage was not presented.

No relevant evidence of damage, that could compromise the films for dosimetry was shown up to an absorbed X-ray dose of 8.0 Gy.

For the same irradiation conditions, there is some indication that FEP film is less resistant than PFA. The analyses used were adequate to detect chemical changes in the film structures.

ACKNOWLEDGMENTS

The present work was supported by grants from CAPES/CNPq (Brazil) and AAMURI (USA).

REFERENCES

1. K. K. Groh, J. R. Gaier, R. L. Hall, M. P. Espe, D. R. Cato, J. K. Sutter, D. A. Scheimank, *High Perform. Polym*, **12:83,** 104 (2000).
2. J. A. Denver, K. K. Groh, A. J. Towsed, L. L. Wang, *AIII paper*, **0895:1** (1998).
3. M. A Parada., A. De Almeida, *Nuclear Instruments and Methods In Physics Research B - Beam Interactions with Materials and Atoms.*, **191:1,** 820 (2002).
4. G. M. Sessler, "*Introdution and physical principles of electrets*", New York, Spring-Verlag, (1980).

5. N. C. da Silva et al., *"X-ray image produced on an electret film"*, in: 9th International Symposium on Electrets (ISE9), (IEEE, China (1996)), pp. 1067-1072.

6. L. N. Rodrigues, *Um novo dosímetro de eletreto para radioterapia*, Master Thesis, IFSC - USP, Brazil, (1985).

7. M. A. Parada et al. *"Charge Density Limits in Polymer Film Electrets"*, ISE12 proceedings from IEEE dielectrics and Electrical Insulation Society (2005), pp. 304-306.

8. C. R. Brundle, C. A. Evans Jr., S. Wilson, *"Encyclopedia of Materials Characterization: surfaces, interfaces, thin films"*, Butterworth-Heinemann, Boston, 1992.

9. M. A. Parada et al, *"Fluoropolymers Studies for Radiation Dosimetry"*, Brazilian Journal of Physics, **34 (3A),** 948 (2004).

AUTHOR INDEX

SUBJECT INDEX

CPSIA information can be obtained at www.ICGtesting.com
Printed in the USA
LVOW06s1007220514

386805LV00011B/306/P